Verification and Control of Hybrid Systems

Paulo Tabuada

Verification and Control of Hybrid Systems

A Symbolic Approach

Foreword by Rajeev Alur

Paulo Tabuada
Department of Electrical Engineering
University of California, Los Angeles
66-147F, Engineering Bldg. IV
Los Angeles, CA 90095-1594
USA
tabuada@ee.ucla.edu

ISBN 978-1-4419-5498-5 e-ISBN 978-1-4419-0224-5
DOI 10.1007/978-1-4419-0224-5
Springer Dordrecht Heidelberg London New York

Printed on acid-free paper

Springer is part of Springer Science+Business Media (www.springer.com)

This book is dedicated to my father.

Stricken again, by the bitterness of life, while this book was being written.

Foreword

It is my great pleasure to write this foreword to this remarkable monograph on hybrid systems by Paulo Tabuada. A hybrid system combines the state-machine models of discrete control with differential-equation models of continuous dynamics. Traditionally, state machines have been used extensively for modeling and analysis in computer science, and dynamical systems are studied by researchers in control theory. The embedded applications, consisting of discrete software reacting to continuous environment, have motivated the study of the problems at the boundary of the two fields, leading to the growing and exciting field of hybrid systems. This research discipline brings together researchers from software engineering and formal methods, control theory, and applications such as robotics and systems biology.

The results at the computer science end were motivated by a desire to generalize the symbolic methods that had proved to be so successful in modeling and analyzing digital hardware. Early results focused on models such as *timed automata* and *linear hybrid automata* that augmented finite-state control with simple dynamics. At the other spectrum, control theorists developed ways for generalizing analysis techniques for linear and non-linear systems to switched counterparts. Over the years, a wealth of results have emerged at the intersection, but more importantly, the cross-fertilization of ideas have led to new insights and analysis techniques. For example, the theory of *property-preserving abstractions*, studied extensively in the formal methods literature, has led to computational techniques for reducing the dimensionality of linear control systems; the concepts of *barrier certificates* and *Lyapunov functions* from dynamical systems theory have led to effective tools for verification of safety and liveness properties of hybrid systems.

The rapidly growing and interdisciplinary research demands that researchers must understand concepts and tools from both computer science and control theory. This is undoubtedly a daunting task, and this is where this book can come to the rescue. It covers a wide range of foundational topics in verification and control of hybrid systems. More impressively, it thematically weaves them together using the recurring theme of *simulation* relations.

The notion of a simulation or a bisimulation relation between two descriptions of a system, possibly at different levels of detail, has proved to be an important concept, both in theory and practice, for modeling and verification of discrete systems. Using richer concepts such as *alternating* relations and *approximate* relations, this book develops the foundations for hybrid control systems.

Paulo is ideally suited for the challenge of explaining the foundations of hybrid systems. Having interacted with him for many years, I am well aware of the breadth of his technical expertise that spans across topics in formal verification and control theory. He has made numerous important contributions to the field, particularly to foundations and tools for property-preserving abstractions of hybrid control systems. The exposition is technically rigorous, and uses a number of well-chosen illustrations to explain the concepts.

There are a number of conferences on hybrid systems. It is also an integral component of the newly emerging and broader discipline of *cyber-physical systems*. There are new challenges, and research opportunities, related to, for instance, robustness in presence of uncertainties and design of networked control systems. I recommend this book highly to students and researchers who want to tackle these challenges. Studying this book will be an excellent way to understand the mathematical foundations of this vibrant field.

University of Pennsylvania Rajeev Alur

Preface

Hybrid systems arose more than 15 years ago in a bold attempt to yoke together computer science and control theory in the context of, what are now called, cyber-physical systems. Although it is still early to give an unified account of hybrid systems research, certain conceptual similarities, between different results developed by different researchers, have recently come into view. This book aims at highlighting these similarities by providing a systematic exposition of several key verification and control synthesis results.

The guiding concept used in this book is the notion of bisimulation. To understand how it winds through hybrid systems research, a digression into digital control and timed automata is in order. Before hybrid systems, the existing paradigm for computer controlled systems was digital control. Under this paradigm, a computer interfaces with the physical world through an Analog-to-Digital (A/D) converter, transforming measured physical quantities into digital format, and through a Digital-to-Analog (D/A) converter, transforming digital control commands into analog signals. The enabling result of this paradigm states that when the continuous dynamics is described by a controlled linear differential equation, the combination of the continuous dynamics with certain classes of D/A and A/D converters can be represented by a discrete-time linear control system. Consequently, discrete-time linear control systems were widely used as the abstraction of choice for computer controlled physical systems. Although powerful, the digital control paradigm provides no support to study the interaction of software with the physical world. The first models for hybrid systems extended the digital control paradigm by modeling the software as a finite-state machine interfacing with the physical world, described by a controlled differential equation, through A/D and D/A converters. This model for hybrid systems later evolved into hybrid automata but no enabling representation result, such as the one for digital control, was known. The desired representation result had already been partially given by Alur and Dill's work on timed automata. In the 1990 paper [AD90], Alur and Dill showed that timed automata, a special class of hybrid automata, could be represented by a finite-state symbolic model, *i.e.*, a model with finitely many states where each state or symbol represents infinitely many states of the timed automaton. Essential to their results was the observation that the notion of bisimulation was versatile enough to establish a formal equivalence between timed automata and finite-state symbolic models. Bisimulation, originally introduced by Park [Par81] and Milner [Mil89] as a notion of equivalence between software processes, provided the motto for extensions of Alur and Dill's pioneering work to other classes of hybrid systems. Although the initial effort was on representation results for the verification of hybrid systems, later results showed that symbolic models could also be constructed for control design. A further twist in this research stream occurred recently when it was recognized that bisimulation could be generalized to approximate bisimulation with the purpose of further enlarging the class of hybrid systems admitting symbolic models.

The excursion into hybrid systems research, offered in this book, is divided into four parts. The first part presents basic concepts centered around a notion of system that is general enough to describe finite-state machines, differential equations, and hybrid systems. However, a system, by itself, is not very interesting. More interesting are the ways in which systems relate to other systems. Two such relationships are presented in Part II: behavioral inclusion/equivalence and simulation/bisimulation. These relationships are then used to study verification and control synthesis problems for finite-state systems. Only a flavor of the existing results is provided since the focus of the book are the infinite-state (hybrid) systems discussed in Part III and Part IV. By drawing inspiration from timed automata, several classes of hybrid systems with richer continuous dynamics are shown to be related to finite-state symbolic systems in Part III. Once such (bi)simulation relations are established, verification and control synthesis problems can be immediately solved by resorting to the techniques described in Part II for finite-state systems. The same strategy is followed in Part IV by generalizing (bi)simulation relationships to approximate (bi)simulation relationships that can be used for wider classes of hybrid systems.

The choice of results presented in this book is admittedly biased by my view of hybrid systems and my own research interests. Moreover, I confess that the guiding concept of bisimulation warrants a longer route visiting many other important results. The decision not to include such topics was difficult to make, but including them would have required more time than I could give to this project at this stage. In addition to the choice of topics, I faced another challenge: to make the book accessible and interesting to both computer scientists and control theorists. On the one hand, the computer scientist will certainly find Part II to be a very narrow account of formal verification and Part III to treat timed automata very superficially. On the other hand, the control theorist will be intrigued with the notion of system used in this book, yet disappointed with the relegation of nonlinear systems to several cursory sections. Nonetheless, I hope the readers, independently of their technical background, will find the results interesting enough to consult the specialized literature for the missing details.

It is a great pleasure to acknowledge the influence of Rajeev Alur and George J. Pappas that helped shape my view of hybrid systems. Some of the results in Part IV were developed in collaboration with Giordano Pola and Antoine Girard and for that, I am very grateful. My students and postdocs, especially Manuel Mazo Jr. and Ramkrishna Pasumarthy, also deserve a special word of acknowledgement for carefully reading different versions of this book and providing me with valuable feedback. This book would not have been possible without the scholarly environment created by my colleagues at the Electrical Engineering Department of the University of California at Los Angeles, and without the support of Dr. Helen Gill and Dr. Radhakishan Baheti from the National Science Foundation. My final words of acknowledgment go to my brother, who showed me how to persevere and prosper among brilliant minds, to my parents, for being a constant source of support, and to my wife, to whom I now have to repay the countless nights and weekends spent in the writing of this book.

Los Angeles and Lisboa *Paulo Tabuada*

Contents

Part I Basic concepts

Part II Finite systems

Part I

Basic concepts

1

Systems

The word system is used in this book as a synonym for *mathematical model of a dynamical phenomenon.* Since different problems may require different models of the same phenomenon, we need a versatile notion of system that can be equipped with relationships explaining how different systems can be related. The purpose of this chapter is to provide one such notion and to illustrate its use in different contexts.

Notation

For a set Z, $1_Z : Z \to Z$ denotes the identity map on Z defined by $1_Z(z) = z$ for every $z \in Z$. Given a map $f : Z \to W$ and a set $K \subseteq Z$, $f(K)$ denotes the subset of W defined by $f(K) = \{w \in W \mid w = f(k) \text{ for some } k \in K\}$ while $f|_K : K \to W$ describes the restriction of f to K defined by $f|_K(k) = f(k)$ for every $k \in K$. The projection map taking $(x_a, x_b, u_a, u_b) \in X_a \times X_b \times U_a \times U_b$ to $(x_a, x_b) \in X_a \times X_b$ is denoted by π_X.

1.1 System definition

Among the many different mathematical models used to describe dynamical phenomena we are especially interested in models with states belonging to finite sets, infinite sets, and combinations thereof. By a finite-state system we mean a system described by finitely many states. The finite-state machines used to model digital circuits are one such example. We also consider infinite-state systems described by difference or differential equations with solutions evolving in infinite sets such as $\mathbb{R}^n$. Hybrid systems, combining aspects of finite-state and infinite-state systems, consist of another class of systems that can be described by the notion of system adopted in this book.

P. Tabuada, *Verification and Control of Hybrid Systems: A Symbolic Approach*,
DOI: 10.1007/978-1-4419-0224-5_1,

Definition 1.1 (System). *A system* S *is a sextuple* $(X, X_0, U, \longrightarrow, Y, H)$ *consisting of:*

- *a set of* states X*;*
- *a set of* initial states $X_0 \subseteq X$*;*
- *a set of* inputs U*;*
- *a* transition relation $\longrightarrow \subseteq X \times U \times X$*;*
- *a set of* outputs Y*;*
- *an* output map $H : X \to Y$.

States in X are regarded as internal to the system whereas outputs are externally visible. The set of initial states may be a proper subset $X_0 \subset X$, a fixed initial state $x_0 \in X$, or the whole set of states $X_0 = X$. A system is called *finite-state* if X is a finite set. A system that is not finite-state is called *infinite-state*. Systems described by differential equations are examples of infinite-state systems. The adjectives finite and infinite always qualify the state set of a system whereas the adjectives *discrete* and *continuous* are used to qualify time. Even though a state cannot[1] be qualified as finite or infinite, we shall abuse language and call a state finite or infinite when the corresponding state set is finite or infinite, respectively. This abuse of language will be extremely useful throughout the book. The relationship between finite-state and infinite-state systems is an important topic that is considered in great detail in this book.

The evolution of a system is captured by the transition relation. A transition $(x, u, x') \in \longrightarrow$ is, throughout the book, denoted by $x \xrightarrow{u} x'$. For such a transition, state x' is called a *u-successor*, or simply *successor*, of state x. Similarly, x is called a *u-predecessor*, or *predecessor*, of state x'. Note that, since $\longrightarrow \subseteq X \times U \times X$ is a relation, for any state and any input $u \in U$ there may be: no u-successors, one u-successor, or many u-successors. For conciseness, we denote the set of u-successors of a state x by $\mathrm{Post}_u(x)$. Since $\mathrm{Post}_u(x)$ may be empty, we denote by $U(x)$ the set of inputs $u \in U$ for which $\mathrm{Post}_u(x)$ is nonempty. As discussed in later chapters, the semantics of the elements in U depends on the problem being solved. Inputs in U can represent choices to be made by a controller, choices to be made by the environment, or they can simply describe the passage of time.

Example 1.2. Finite-state systems naturally arise as models of a variety of man-made phenomena. In addition to being completely defined by the data described in Definition 1.1, they also admit a graphical representation that is very useful. States are represented by circles and transitions are represented by arrows between states. Initial states are distinguished by being the target of a sourceless arrow. Each circle is labeled by the state (top half) and the

[1] A state is simply an element of the set of states X. Therefore, unless additional structure is imposed on X, the expressions "infinite state" and "finite state" have no defined meaning.

corresponding output (bottom half), and each arrow is labeled by the input. The graphical representation of the finite-state system defined by the data:

$$X = \{x_0, x_1, x_2, x_3\}, \quad X_0 = \{x_0, x_2\}, \quad U = \{u_0, u_1\}, \tag{1.1}$$

$$\begin{aligned} \longrightarrow = \{&(x_0, u_0, x_1), (x_0, u_1, x_2), (x_1, u_0, x_1), \\ &(x_1, u_0, x_3), (x_2, u_1, x_3), (x_3, u_1, x_1)\}, \end{aligned} \tag{1.2}$$

$$Y = \{y_0, y_1, y_2\}, \tag{1.3}$$

$$H(x_0) = y_0, \quad H(x_1) = y_0, \quad H(x_2) = y_1, \quad H(x_3) = y_2, \tag{1.4}$$

is displayed in Figure 1.1. $\triangleleft$

A system is called *blocking* if there is a state $x \in X$ from which no further transitions are possible, *i.e.*, x has no u-successors for any $u \in U$. This can also be expressed as $U(x) = \varnothing$. A system is called *nonblocking* if the set of successors of every $x \in X$ is nonempty. An equivalent characterization is $U(x) \neq \varnothing$ for every $x \in X$.

A system is called *deterministic* if for any state $x \in X$ and any input $u \in U$, $x \xrightarrow{u} x'$ and $x \xrightarrow{u} x''$ imply $x' = x''$. Therefore, a system is deterministic if given any state $x \in X$ and any input $u \in U$, there exists at most one u-successor (there may be none). A system is *output deterministic* if: $H|_{X_0}$ is injective; and for any state $x \in X$ and any inputs $u, u' \in U$, $x \xrightarrow{u} x'$ and $x \xrightarrow{u'} x''$ with $H(x') = H(x'')$ imply $x' = x''$. For output deterministic systems, different successors of a state always have different outputs.

A system is called *nondeterministic* if it is not deterministic. Hence for a nondeterministic system it is possible for a state to have two (or possibly more) distinct u-successors.

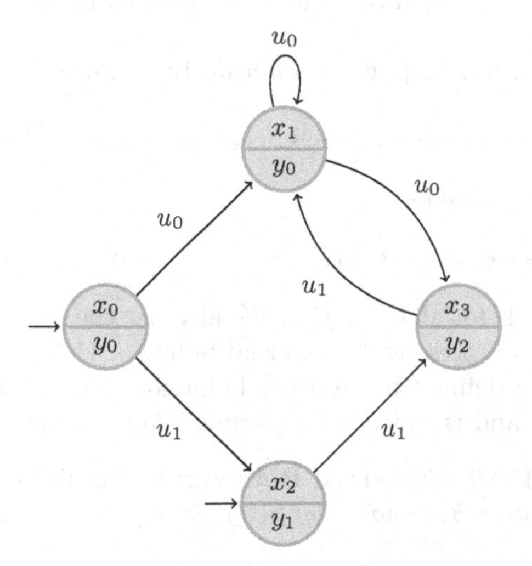

Fig. 1.1. Graphical representation of the finite-state system defined by (1.1) through (1.4).

One can easily see that the system represented in Figure 1.1 is nonblocking since every state has an outgoing transition. It is also nondeterministic as there are two u_0-successors of the state x_1, namely x_1 and x_3. Albeit not being deterministic this system is output deterministic.

To simplify notation we also denote a system $S = (X, X_0, U, \longrightarrow, Y, H)$ by the quintuple $S = (X, U, \longrightarrow, Y, H)$ when $X_0 = X$, by the quadruple $S = (X, X_0, U, \longrightarrow)$ when $Y = X$ and $H = 1_X$, or by the triple $S = (X, U, \longrightarrow)$ when $X_0 = X = Y$ and $H = 1_X$.

1.2 System behavior

Given any state $x \in X$, a *finite internal behavior* generated from x is a finite sequence of transitions:

$$x_0 \xrightarrow{u_0} x_1 \xrightarrow{u_1} x_2 \xrightarrow{u_2} \dots \xrightarrow{u_{n-2}} x_{n-1} \xrightarrow{u_{n-1}} x_n$$

such that $x_0 = x$ and $x_i \xrightarrow{u_i} x_{i+1}$ for all $0 \leq i < n$. A state $x \in X$ can also be seen as a behavior comprising zero transitions in which case $n = 0$. An internal behavior generated from x is *initialized* if $x \in X_0$.

In some cases, a finite internal behavior can be extended to an infinite internal behavior. An *infinite internal behavior* generated from x is an infinite sequence:

$$x_0 \xrightarrow{u_0} x_1 \xrightarrow{u_1} x_2 \xrightarrow{u_2} x_3 \xrightarrow{u_3} \dots$$

that satisfies $x_0 = x$ and $x_i \xrightarrow{u_i} x_{i+1}$ for all $i \in \mathbb{N}_0$. An infinite internal behavior generated from x is called *initialized* if $x \in X_0$. In nonblocking systems, every finite internal behavior can be extended to an infinite internal behavior.

Through the output map, every internal behavior:

$$x_0 \xrightarrow{u_0} x_1 \xrightarrow{u_1} x_2 \xrightarrow{u_2} \dots \xrightarrow{u_{n-2}} x_{n-1} \xrightarrow{u_{n-1}} x_n$$

defines an external behavior:

$$y_0 \longrightarrow y_1 \longrightarrow y_2 \longrightarrow \dots \longrightarrow y_{n-1} \longrightarrow y_n \tag{1.5}$$

with $H(x_i) = y_i \in Y$ for all $0 \leq i \leq n$. We also use the more succinct notation $\mathsf{y} = y_0 y_1 y_2 \dots y_n$ to represent the external behavior (1.5). The set of external behaviors that are defined by internal behaviors generated from state x is denoted by $\mathcal{B}_x(S)$ and is called the *external behavior* from state x.

Definition 1.3 (Finite External Behavior). *The* finite external behavior *generated by a system S, denoted by $\mathcal{B}(S)$, is defined by:*

$$\mathcal{B}(S) = \bigcup_{x \in X_0} \mathcal{B}_x(S).$$

For output deterministic systems any finite external behavior y determines uniquely the corresponding internal behavior. This can easily be shown by induction. Given an external behavior $\mathsf{y} = y_0 y_1 y_2 \dots y_n$ we can recover the corresponding initial state x_0 since $H|_{X_0}$ is injective. Then, we consider all the successors x_1 of x_0 satisfying $H(x_1) = y_1$. If there is more than one successor, say x_1 and x_1', we have $x_0 \xrightarrow{u_1} x_1$ and $x_0 \xrightarrow{u_1'} x_1'$ with $H(x_1) = H(x_1')$. It follows by output determinism that $x_1 = x_1'$ and x_1 is uniquely determined. Applying the same argument to the successors of x_1 we can uniquely recover x_2 and so on.

An infinite internal behavior from x:

$$x_0 \xrightarrow{u_0} x_1 \xrightarrow{u_1} x_2 \xrightarrow{u_2} x_3 \xrightarrow{u_3} \dots$$

defines an infinite external behavior:

$$y_0 \longrightarrow y_1 \longrightarrow y_2 \longrightarrow y_3 \longrightarrow \dots \tag{1.6}$$

corresponding to the infinite sequence of outputs with $H(x_i) = y_i$ for all $i \in \mathbb{N}_0$. The infinite external behavior (1.6) can also be succinctly denoted by $\mathsf{y} = y_0 y_1 y_2 y_3 \dots$. The set of all infinite external behaviors that are generated from x is denoted by $\mathcal{B}^{\omega}_x(S)$ and called the *infinite external behavior* from state x.

Definition 1.4 (Infinite External Behavior). *The* infinite external behavior *generated by a system S, denoted by $\mathcal{B}^{\omega}(S)$, is defined by:*

$$\mathcal{B}^{\omega}(S) = \bigcup_{x \in X_0} \mathcal{B}^{\omega}_x(S).$$

Infinite behaviors describe the nonterminating interaction of a system with other systems and the environment. They are thus adequate to model the operation of reactive systems, such as embedded controllers, that must operate without interruption for arbitrarily long periods of time. For this reason, we focus mostly on infinite behaviors and drop the adjective infinite whenever clear from the context.

If a system S is non-blocking, then $\mathcal{B}^{\omega}(S)$ is nonempty. However, $\mathcal{B}^{\omega}(S)$ may be nonempty even if S is a blocking system. Figure 1.2 displays one such example where the infinite external behavior aaaaa... belongs to $\mathcal{B}^{\omega}(S)$ although S is blocking since the state x_1 has no successors.

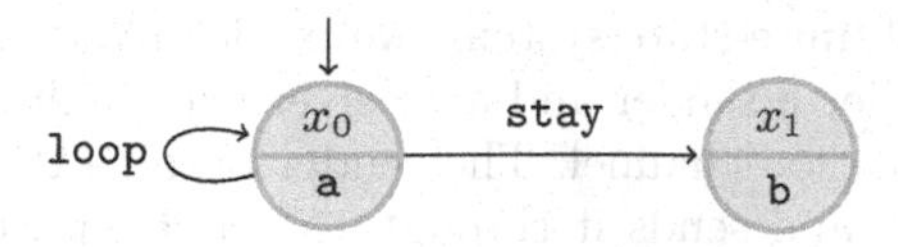

Fig. 1.2. Graphical representation of a blocking finite-state system with nonempty infinite external behavior.

In Chapter 4 we discuss in further detail the relation between $\mathcal{B}(S)$, $\mathcal{B}^\omega(S)$, and the blocking/nonblocking properties of S. For now, we veer to the relation between inputs and outputs. In many situations the inputs provide valuable information that is not directly captured by the internal or external behavior of a system. This can be easily remedied by suitably extending the state set. Starting from a system $S = (X, X_0, U, \longrightarrow, Y, H)$ we can construct a new system $S_o = (X_o, X_{o0}, U_o, \underset{o}{\longrightarrow}, Y_o, H_o)$ with:

- $X_o = X \times U$;
- $X_{o0} = X_0 \times \{*\}$ for some element $* \notin U$;
- $U_o = U \cup \{*\}$;
- $(x, u) \overset{u'}{\underset{o}{\longrightarrow}} (x', u')$ in S_o if $x \overset{u'}{\longrightarrow} x'$ in S;
- $Y_o = Y \times U$;
- $H_o(x, u) = (H(x), u)$.

An infinite external behavior of S_o is of the form:

$$(y_0, *) \longrightarrow (y_1, u_0) \longrightarrow (y_2, u_1) \longrightarrow (y_3, u_2) \longrightarrow \ldots$$

thus containing not only the infinite external behavior of S:

$$y_0 \longrightarrow y_1 \longrightarrow y_2 \longrightarrow y_3 \longrightarrow \ldots \tag{1.7}$$

but also the sequence of inputs $u_0 u_1 u_2 \ldots$ used to generated (1.7) in S. Hence, the output set Y and output map H can be designed to make externally visible the aspects (inputs and states) of a system that are considered relevant for the verification or control problem being solved.

1.3 Examples

The examples that follow illustrate the versatility of the notion of system adopted in this book. The results presented in Part II, III, and IV apply not only to the examples in this section but also to many other examples that can be suitably described by the adopted notion of system.

1.3.1 Finite-state systems

Communication protocol

As a first example of finite-state systems we model a very simple communication protocol. Consider a sender and a receiver exchanging messages over an unreliable communication channel. The sender obtains the data to be transmitted from a buffer and sends it through the channel. Since the channel is unreliable, the sender waits for a confirmation message from the receiver. If the confirmation message acknowledges a correct reception, new data is fetched

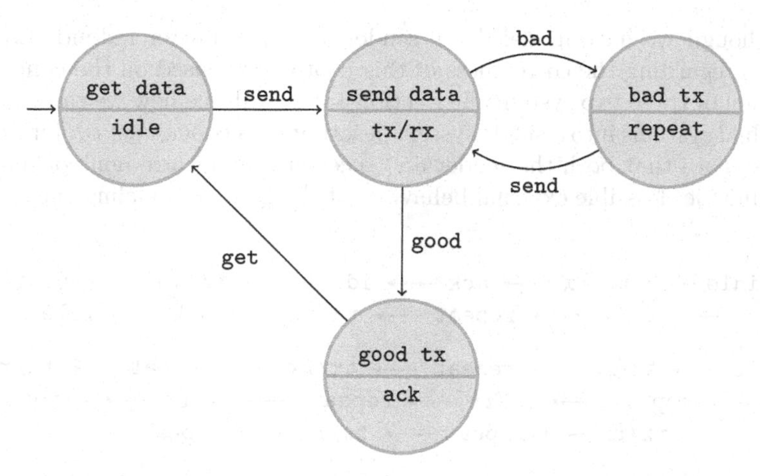

Fig. 1.3. Graphical representation of the finite-state system modeling the sender.

and transmitted. When the confirmation message acknowledges an incorrect reception, the previous message is resent. The behavior of the sender can be described by the finite-state system in Figure 1.3.

The communication channel can either deliver the sent message without errors or deliver a corrupted version of the sent message. The communication channel is not independently modeled but is integrated in the model of the sender and the model of the receiver.

The receiver sends a confirmation message acknowledging the reception of an error free message or a repeat request when the received message is corrupted. The receiver is described by the finite-state system in Figure 1.4.

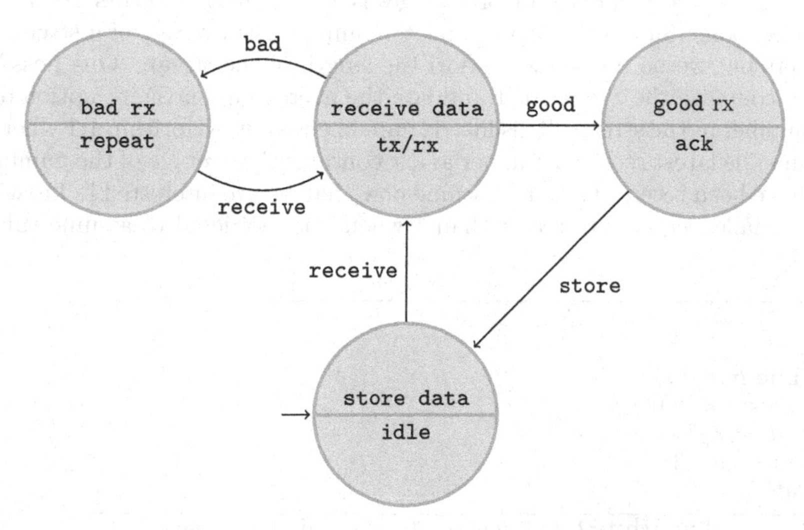

Fig. 1.4. Graphical representation of the finite-state system modeling the receiver.

Although we have modeled the sender and transmitter individually, any analysis regarding the correctness of this protocol is based on the concurrent execution of these two systems. In Section 1.4 we will see how we can compose individual systems into a single system describing its concurrent evolution. For now, we note that both the sender and receiver systems are nonblocking and deterministic. Possible external behaviors of the system modeling the receiver are:

$$\texttt{idle} \longrightarrow \texttt{tx/rx} \longrightarrow \texttt{ack} \longrightarrow \texttt{idle} \longrightarrow \texttt{tx/rx} \longrightarrow \texttt{repeat} \longrightarrow \texttt{tx/rx} \longrightarrow \texttt{repeat} \longrightarrow \texttt{tx/rx} \longrightarrow \texttt{ack} \longrightarrow \texttt{idle}$$

$$\texttt{idle} \longrightarrow \texttt{tx/rx} \longrightarrow \texttt{repeat} \longrightarrow \texttt{tx/rx} \longrightarrow \texttt{repeat} \longrightarrow \texttt{tx/rx} \longrightarrow \texttt{repeat} \longrightarrow \texttt{tx/rx} \longrightarrow \texttt{repeat} \longrightarrow \texttt{tx/rx} \longrightarrow \texttt{repeat} \longrightarrow \texttt{tx/rx} \longrightarrow \texttt{repeat} \longrightarrow \texttt{tx/rx} \longrightarrow \texttt{repeat} \longrightarrow \ldots$$

$$\texttt{idle} \longrightarrow \texttt{tx/rx} \longrightarrow \texttt{repeat} \longrightarrow \texttt{tx/rx} \longrightarrow \texttt{ack} \longrightarrow \texttt{idle} \longrightarrow \texttt{tx/rx} \longrightarrow \texttt{repeat} \longrightarrow \texttt{tx/rx} \longrightarrow \texttt{ack} \longrightarrow \texttt{idle} \longrightarrow \texttt{tx/rx}$$

Software

Finite-state systems can also be used to model software. Intuitively, we can regard the memory contents of a computational device as its state and the software as a description of how memory contents change over time. Since we can only store finitely many bits of information in the memory of a computing system, the set of states is necessarily finite.

Constructing finite-state models of software is a challenging task that we shall not address in this book. Instead, we provide a very simple example.

Example 1.5. The computation of averages is a problem that occurs frequently in applications. Suppose that we want to compute the average of a stream of numbers but we do not know a priori the length of the stream. One possible way to compute the average is to update the average upon the reception of a new number in the stream. This idea is implemented by Algorithm 1.1 where y contains the latest received number and x contains the average of the numbers that have been received so far. Assume now that we are interested in knowing if x is smaller, equal, or greater than 1 when y is restricted to assume values

x:=0;
n:=0;
while *true* **do**
  y:=read(input);
  $x := x\frac{n}{n+1} + y\frac{1}{n+1}$;
  $n := n + 1$;
end

Algorithm 1.1: Average of a stream of numbers.

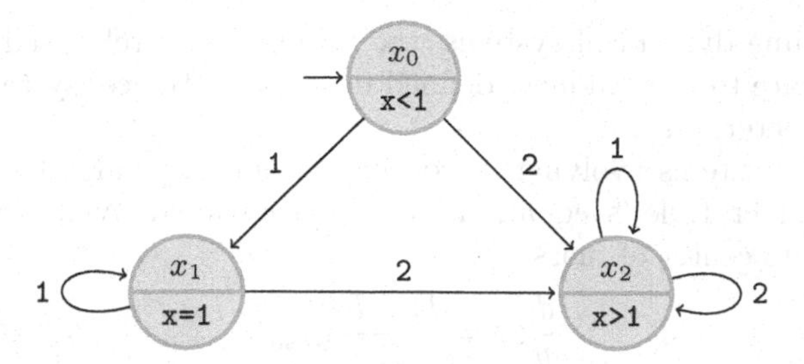

Fig. 1.5. Finite-state system describing if $x < 1$, $x = 1$, or $x > 1$ when $y \in \{1, 2\}$ and x evolves according to Algorithm 1.1.

in the set $\{1, 2\}$. One possible finite-state model capturing the dynamics of x is represented in Figure 1.5. Every execution of the while loop is captured by a transition of the system. $\lhd$

1.3.2 Infinite-state systems

Dynamical systems

Paul Samuelson introduced in 1939 the following model for the evolution of the national income:

$$\begin{aligned}
y(n) &= c(n) + i(n) + g(n)\\
c(n+1) &= \alpha y(n)\\
i(n+1) &= \beta(c(n+1) - c(n))
\end{aligned}$$

where α and γ are parameters and y denotes the national income that is formed by three different kinds of expenditures: consumption, denoted by c; investment, denoted by i; and government expenditure, denoted by g. Eliminating y we obtain:

$$\begin{aligned}
c(n+1) &= \alpha\big(c(n) + i(n) + g(n)\big)\\
i(n+1) &= \beta\alpha\big(c(n) + i(n) + g(n)\big) - \beta c(n).
\end{aligned}$$

If we assume that g is a fixed value, the preceding model defines the system S with:

- $X = (\mathbb{R}_0^+)^2$;
- $X_0 = X$;
- $U = \{*\}$;
- $(c, i) \xrightarrow{*} (c', i')$ iff $c' = \alpha(c + i + g)$ and $i' = \beta\alpha(c + i + g) - \beta c$;
- $Y = \mathbb{R}_0^+$;
- $H(c, i) = c + i + g$.

For discrete-time dynamical systems, the input $*$ is merely used to advance the current state to the uniquely defined next state. Hence, system S is nonblocking and deterministic.

Dynamical systems evolving in continuous-time can also be modeled as systems. Consider Euler's equations for the rotational dynamics of a rigid body around its center of mass:

$$\frac{d}{dt}\xi_1 = \frac{I_2 - I_3}{I_1}\xi_2\xi_3 \tag{1.8}$$

$$\frac{d}{dt}\xi_2 = \frac{I_3 - I_1}{I_2}\xi_3\xi_1 \tag{1.9}$$

$$\frac{d}{dt}\xi_3 = \frac{I_1 - I_2}{I_3}\xi_1\xi_2 \tag{1.10}$$

where $\xi = (\xi_1, \xi_2, \xi_3)$ is the body angular velocity and $I_1, I_2, I_3 \in \mathbb{R}$ are the principal moments of inertia. A solution to these equations with initial condition $x_0 = (x_{10}, x_{20}, x_{30})$ is a continuously differentiable curve $\xi :]a, b[\rightarrow \mathbb{R}^3$ with $a < 0 < b$, satisfying $\xi(0) = x_0$ and (1.8) through (1.10) for all $t \in]a, b[$. We can thus model the preceding continuous-time dynamical system as a system S with:

- $X = \mathbb{R}^3$;
- $X_0 = X$;
- $U = \mathbb{R}_0^+$;
- $x \xrightarrow{\tau} x'$ if there exists a solution ξ to equations (1.8) through (1.10) satisfying $\xi(0) = x$ and $\xi(\tau) = x'$;
- $Y = X$;
- $H = 1_X$.

The preceding construction encodes in the transition relation of S all the information contained in the solution ξ since there exists a transition $x \xrightarrow{\tau} x'$ in S iff $\xi(0) = x$ and $\xi(\tau) = x'$. Although this is natural from a mathematical point of view, some readers may feel unsettled with this modeling choice: by labeling the transitions with time we are formally treating time as the input of S. If one thinks of dynamics as change in space corresponding to change in time, then one can easily reconcile oneself with the idea of treating time as an input. Although one can conceive the existence of a uniform notion of time across all systems, different interactions with a system may require different time inputs. As an example, consider two different digital platforms measuring the output of system S. If platform A has a clock cycle of 2 time units and platform B has a clock cycle of 3 time units, then platform A feeds time inputs from the set $\{0, 2, 4, 6, \ldots\}$ into S while platform B feeds time inputs from the set $\{0, 3, 6, 9, \ldots\}$ into S. Treating time as the input also has the added benefit of rendering S nonblocking and deterministic as an immediate consequence of existence and uniqueness of solutions for smooth differential equations. In Chapter 7 and Chapter 10 we discuss different ways in which dynamical systems can be modeled as systems.

Control systems

The national income model described in the previous section can be regarded as a control system if one assumes that the government expenditure at time $n+1$ can be chosen by the government. In this case we obtain the following equations:

$$\begin{aligned} c(n+1) &= \alpha\big(c(n)+i(n)+g(n)\big) \\ i(n+1) &= \beta\alpha\big(c(n)+i(n)+g(n)\big) - \beta c(n) \\ g(n+1) &= u(n) \end{aligned}$$

where u is regarded as an input to be chosen by the government and belonging to the set $[0, D]$ for some maximal expenditure $D \in \mathbb{R}^+$. This model defines a system with:

- $X = (\mathbb{R}_0^+)^3$;
- $X_0 = X$;
- $U = [0, D]$;
- $(c, i, g) \xrightarrow{u} (c', i', g')$ iff $c' = \alpha(c + i + g)$, $i' = \beta\alpha(c + i + g) - \beta c$, and $g' = u$;
- $Y = \mathbb{R}_0^+$;
- $H(c, i, g) = c + i + g$.

Control systems in continuous-time can also be described as systems. Consider the motion of a satellite subject to the torque produced by gas jet actuators. Its dynamics can be described by the forced version of Euler's equations:

$$\frac{d}{dt}\xi_1 = \frac{I_2 - I_3}{I_1}\xi_2\xi_3 + v_1 \tag{1.11}$$

$$\frac{d}{dt}\xi_2 = \frac{I_3 - I_1}{I_2}\xi_3\xi_1 + v_2 \tag{1.12}$$

$$\frac{d}{dt}\xi_3 = \frac{I_1 - I_2}{I_3}\xi_1\xi_2 \tag{1.13}$$

where the parameters v_1 and v_2 are inputs describing the effect of the gas jets. A system model for the controlled satellite requires a choice of curves $\mathcal{U}$, from $]a, b[$ to $\mathbb{R}^2$, to be used as inputs. From a mathematical point of view, we need to restrict $\mathcal{U}$ to a class of functions that are regular enough to guarantee existence and uniqueness of solutions for the differential equations (1.11) through (1.13). From an engineering point of view, the curves in $\mathcal{U}$ describe signals that are implemented by physical actuators. Therefore, there are limitations on the curves in $\mathcal{U}$ stemming from the actuator technology. Once the set of input curves $\mathcal{U}$ is chosen, we can describe the dynamics of the controlled satellite by the system S with:

- $X = \mathbb{R}^3$;
- $X_0 = X$;
- $U = \mathcal{U}$;
- $x \xrightarrow{\upsilon} x'$ if there exist curves[2] $\mathcal{U} \ni \upsilon : [0, \tau] \to \mathbb{R}^2$ and $\xi : [0, \tau] \to \mathbb{R}^3$ satisfying equations (1.11) through (1.13) with $\xi(0) = x$ and $\xi(\tau) = x'$;
- $Y = X$;
- $H = 1_X$.

The sedulous reader certainly noticed that system S is also driven by time. An input $\upsilon \in \mathcal{U}$ is a curve $\upsilon : [0, \tau] \to \mathbb{R}^2$ describing not only the contribution of the gas jets to the angular acceleration, but also the *length of time* τ during which such contribution lasts. In particular, we can recover equations (1.8) through (1.10) from equations (1.11) through (1.13) by restricting $\mathcal{U}$ to be the set of all input curves with codomain $\{(0, 0)\} \subset \mathbb{R}^2$. In such case, each input $\upsilon \in \mathcal{U}$ is a constant curve assuming the value $(0, 0)$ on its domain $[0, \tau]$. Hence, we can identify υ with the length τ of its domain, thereby identifying $\mathcal{U}$ with $\mathbb{R}_0^+$, and recovering time as the input of system S modeling the rigid body without gas jets. Different system models for control systems appear in Chapter 8 and in Chapter 11.

1.3.3 Hybrid systems

Hybrid systems combine the characteristics of finite-state systems with those of infinite-state systems. We introduce them through examples.

Real-time scheduling

In many safety-critical applications the execution of software tasks needs to be completed in a timely fashion. Timeliness is typically formulated by equipping each software task with a worst case execution time C, a relative deadline D, and, in the case of periodic tasks, a period T. This formulation entails that a task becomes active every T units of time and its execution must finish no later than D units of time after becoming active. A scheduler is a special task that decides which of the active tasks should be executed by the processor. The decision process takes into account the period, relative deadline, and the fact that the execution may take, in the worst case, C units of time to complete. A set of tasks is said to be schedulable if it can be executed without violating any of the deadlines. We can model a periodic task by the hybrid system in Figure 1.6 where we assume that $C < D < T$ so that the task execution

[2] We use curves υ and ξ defined on closed sets while implicitly assuming the existence of curves $\upsilon' :]a, b[\to \mathbb{R}^2$ and $\xi' :]a, b[\to \mathbb{R}^3$ satisfying $\upsilon = \upsilon'|_{[0,\tau]}$, $\xi = \xi'|_{[0,\tau]}$, and all the additional conditions imposed on υ and ξ.

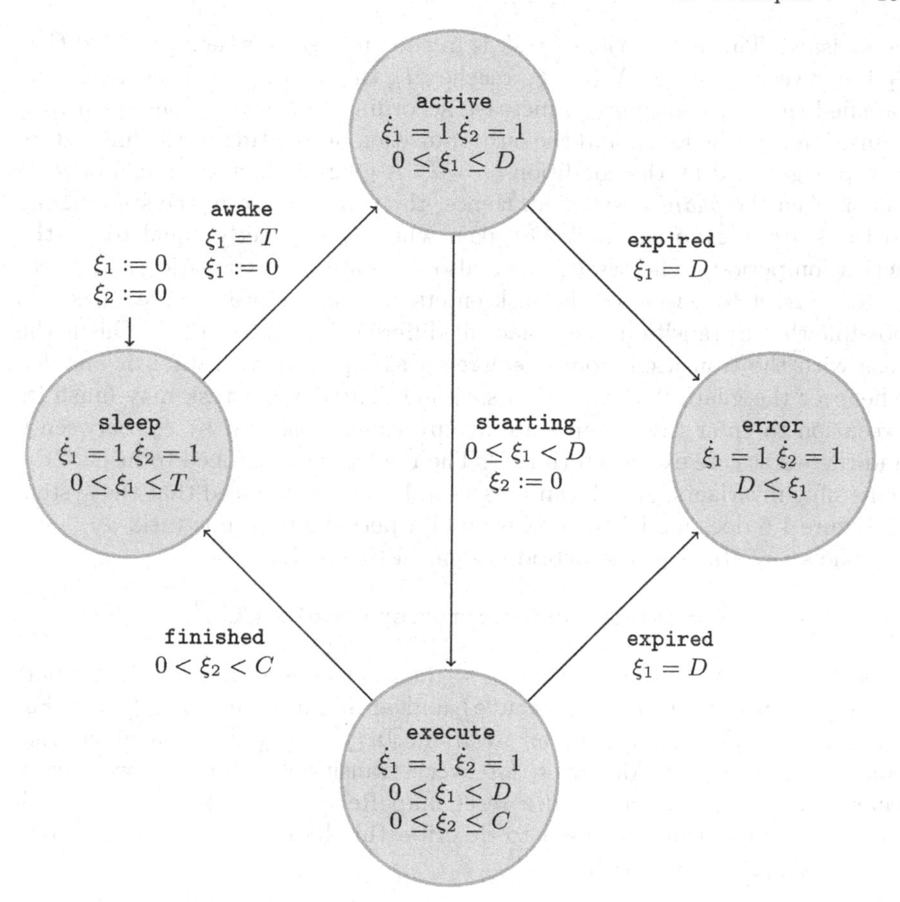

Fig. 1.6. Hybrid system representing a periodic real-time task.

can finish before the relative deadline and before the next activation. We also assume that the task cannot be pre-empted, *i.e.*, once the execution starts the task runs to completion.

There are several new ingredients in Figure 1.6 with respect to the finite-state systems that appeared so far. Firstly, each finite state is decorated with a differential equation which, in this example, is the same for all finite states:

$$\dot{\xi}_1 = 1 \tag{1.14}$$

$$\dot{\xi}_2 = 1. \tag{1.15}$$

The special form of this equation allows us to regard ξ_1 and ξ_2 as clocks measuring the passage of time since $\xi_1(t) = \xi_1(0) + t$ and $\xi_2(t) = \xi_2(0) + t$. Secondly, each finite state is also decorated with an *invariant* set. For example, the finite state `sleep` has $0 \leq \xi_1 \leq T$ as its invariant set. The invariant represents a condition on the solutions of the differential equation that must be satisfied. This means that a task is in the finite state `sleep` provided that ξ_1 is between 0 and T. When ξ_1 reaches T, the invariant can no longer be satisfied since ξ_1 continues to increase according to $\dot{\xi}_1 = 1$. Consequently, a transition must be taken and the only transition originating at the finite state `sleep` is guarded by the condition $\xi_1 = T$. A guarded transition can only be taken when the *guard* is satisfied. Hence, the transition from the state `sleep` to the state `active` can only be taken when ξ_1 is exactly equal to T (the activation period). The *reset* $\xi_1 := 0$ also decorates this transition and forces ξ_1 to be reset to zero once the task enters the finite state `active`. It is also possible that a transition takes place at different instants of time. This is the case with the transition from `execute` to `sleep`. This transition is enabled whenever the guard $0 < \xi_2 < C$ is satisfied. Therefore, a task may finish its execution to enter the `sleep` state at any time, measured by ξ_2, between 0 and the worst case execution time C. The reader is encouraged to inspect the remaining invariants, guards, and resets to become convinced that the system in Figure 1.6 does model the evolution of a periodic real-time task.

The set of states of the hybrid system in Figure 1.6 is:

$$X = \{\mathtt{sleep}, \mathtt{active}, \mathtt{error}, \mathtt{execute}\} \times (\mathbb{R}_0^+)^2.$$

A state $x \in X$ is thus a pair $x = (x_a, x_b)$ consisting of a finite part $x_a \in \{\mathtt{sleep}, \mathtt{active}, \mathtt{error}, \mathtt{execute}\}$ and an infinite part $x_b \in (\mathbb{R}_0^+)^2$. For every finite state x_a we have an invariant $\mathrm{In}_{x_a} \subseteq (\mathbb{R}_0^+)^2$ and a differential equation $\dot{\xi} = f_{x_a}(\xi)$. Moreover, for every transition (x_a, u_a, x_a') we have a guard $\mathrm{Gu}_{(x_a,u_a,x_a')} \subseteq \mathrm{In}_{x_a}$ and a reset map $\mathrm{Re}_{(x_a,u_a,x_a')} : \mathrm{In}_{x_a} \to \mathrm{In}_{x_a'}$. All these new ingredients are used to describe this hybrid system as a system $S = (X, X_0, U, \longrightarrow)$ with:

- $X = \{\mathtt{sleep}, \mathtt{active}, \mathtt{error}, \mathtt{execute}\} \times (\mathbb{R}_0^+)^2$;
- $X_0 = \{(x_a, x_b) \in X \mid x_a = \mathtt{sleep} \ \wedge \ x_b \in \mathrm{In}_{\mathtt{sleep}}\}$;
- $U = \{\mathtt{awake}, \mathtt{expired}, \mathtt{starting}, \mathtt{finished}\} \cup \mathbb{R}_0^+$;
- $(x_a, x_b) \xrightarrow{u} (x_a', x_b')$ if one of the following two conditions holds:
  1. $u \in \mathbb{R}_0^+$, $x_a' = x_a$, and there exists a solution $\xi : [0, u] \to \mathrm{In}_{x_a}$ to the differential equations (1.14) and (1.15) satisfying $\xi(0) = x_b$ and $\xi(u) = x_b'$;
  2. $u \in \{\mathtt{awake}, \mathtt{expired}, \mathtt{starting}, \mathtt{finished}\}$, $x_b' = \mathrm{Re}_{(x_a,u,x_a')}(x_b)$, and $x_b \in \mathrm{Gu}_{(x_a,u,x_a')}$.

The transition relation consists of two different kinds of transitions: continuous flows and discrete transitions. During continuous flows the finite part of the state remains unaltered and the infinite part of the state, whose evolution is prescribed by the differential equations (1.14) and (1.15), remains inside the invariant set. Discrete transitions may change the finite and the infinite part of the state and are conditioned by the corresponding guards.

A boost DC-DC converter

A different example of a hybrid system is the boost DC-DC converter represented in Figure 1.7. If one uses the current i_L through the inductor and the voltage v_C across the capacitor as state variables, a simple application of Kirchoff's laws provides the equations describing the evolution of v_C and i_L. When the switch is in position s_1 we have:

$$\frac{d}{dt}i_L = -\frac{R_L}{L}i_L + \frac{1}{L}v_S \tag{1.16}$$

$$\frac{d}{dt}v_C = -\frac{1}{C}\frac{1}{R_C + R_0}v_C \tag{1.17}$$

and when the switch is in position s_2:

$$\frac{d}{dt}i_L = -\frac{1}{L}\left(R_L + \frac{R_C R_0}{R_C + R_0}\right)i_L - \frac{1}{L}\frac{R_0}{R_C + R_0}v_C + \frac{1}{L}v_S \tag{1.18}$$

$$\frac{d}{dt}v_C = \frac{1}{C}\frac{R_0}{R_C + R_0}i_L - \frac{1}{C}\frac{1}{R_C + R_0}v_C. \tag{1.19}$$

There are two different modes of operation associated with the two different positions for the switch. The dynamics of these two modes of operation can be described by the finite-state system in Figure 1.8 in which the input u_1 takes the system to switch position s_1 and input u_2 takes the system to switch position s_2. However, the finite-state system in Figure 1.8 is not detailed enough to capture the dynamics of i_L and v_C. A more detailed model

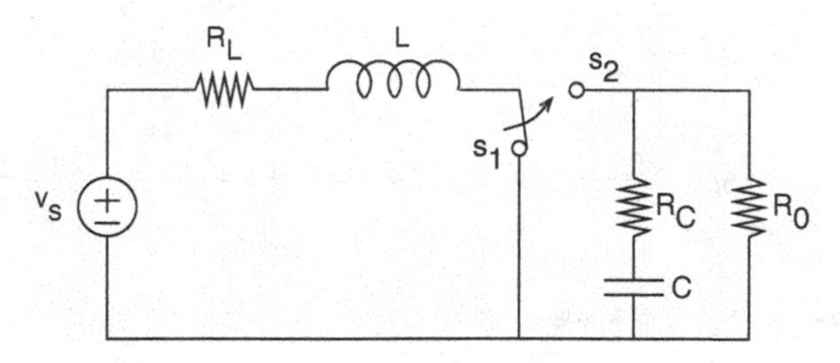

Fig. 1.7. Boost DC-DC converter.

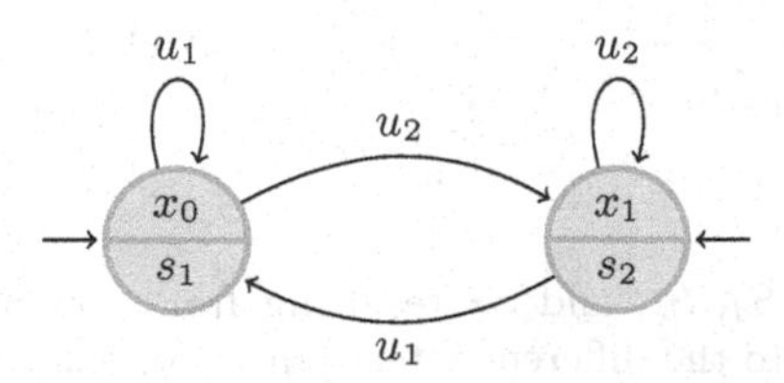

Fig. 1.8. Finite-state system describing the dynamics of the switch in the converter represented in Figure 1.7.

has to combine the finite-state dynamics of the switch with the continuous-time dynamics in equations (1.16) through (1.19). Such hybrid model can be described by the system $S = (X, U, \longrightarrow)$ with:

- $X = \{s_1, s_2\} \times \mathbb{R}^2$;
- $U = \{u_1, u_2\} \cup \mathbb{R}_0^+$;
- $(s, i_L, v_C) \xrightarrow{u} (s', i'_L, v'_C)$ if one of the following four conditions holds:
  1. $u = u_1$, $s' = s_1$, $i_L = i'_L$, and $v_C = v'_C$;
  2. $u = u_2$, $s' = s_2$, $i_L = i'_L$, and $v_C = v'_C$;
  3. $u \in \mathbb{R}_0^+$, $s' = s = s_1$ and there exists a solution $\xi : [0, u] \to \mathbb{R}^2$ to the differential equations (1.16) and (1.17) satisfying $\xi(0) = (i_L, v_C)$ and $\xi(u) = (i'_L, v'_C)$;
  4. $u \in \mathbb{R}_0^+$, $s' = s = s_2$ and there exists a solution $\xi : [0, u] \to \mathbb{R}^2$ to the differential equations (1.18) and (1.19) satisfying $\xi(0) = (i_L, v_C)$ and $\xi(u) = (i'_L, v'_C)$.

For this hybrid system the invariants are $\text{In}_{s_1} = \mathbb{R}^2 = \text{In}_{s_2}$, the guards are $\text{Gu}_{(s_1,u,s_2)} = \mathbb{R}^2 = \text{Gu}_{(s_2,u,s_1)}$, and the reset maps are $\text{Re}_{(s_1,u,s_2)} = 1_{\mathbb{R}^2} = \text{Re}_{(s_2,u,s_1)}$ for any $u \in \{u_1, u_2\}$. The nature of the guards and of the invariants implies that a discrete transition can take place independently of the specific value of the infinite state. This is in stark contrast with the previous example where discrete transitions were influenced by and influenced the continuous-time dynamics. The modeling of hybrid systems as systems is discussed in more detail in Chapter 7.

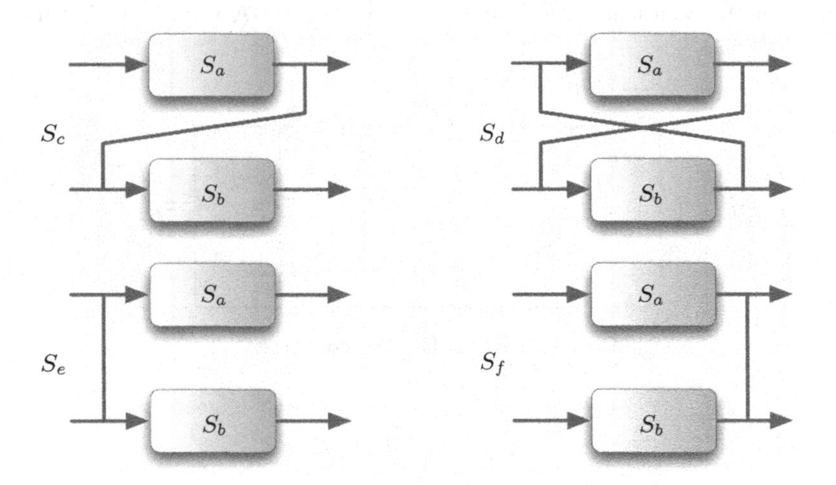

Fig. 1.9. Systems S_c, S_d, S_e, and S_f resulting from composing system S_a with system S_b with respect to the different interconnection relations in Table 1.1.

1.4 Composing systems

Most engineering systems are designed and built by interconnecting simpler and smaller components. This process of constructing larger systems by interconnecting smaller ones can be mathematically described through a composition operation. In this book we consider a versatile notion of composition based on an *interconnection relation* $\mathcal{I} \subseteq X_a \times X_b \times U_a \times U_b$ describing how system S_a interacts with system S_b.

Definition 1.6 (Composition). *Let* $S_a = (X_a, X_{a0}, U_a, \underset{a}{\longrightarrow}, Y_a, H_a)$ *and* $S_b = (X_b, X_{b0}, U_b, \underset{b}{\longrightarrow}, Y_b, H_b)$ *be two systems and let* $\mathcal{I} \subseteq X_a \times X_b \times U_a \times U_b$ *be a relation. The* composition *of* S_a *and* S_b *with interconnection relation* $\mathcal{I}$, *denoted by* $S_a \times_{\mathcal{I}} S_b$, *is the system* $(X_{ab}, X_{ab0}, U_{ab}, \underset{ab}{\longrightarrow}, Y_{ab}, H_{ab})$ *consisting of:*

- $X_{ab} = \pi_X(\mathcal{I})$;
- $X_{ab0} = X_{ab} \cap (X_{a0} \times X_{b0})$;
- $U_{ab} = U_a \times U_b$;
- $(x_a, x_b) \overset{(u_a,u_b)}{\underset{ab}{\longrightarrow}} (x'_a, x'_b)$ *if the following three conditions hold:*
  1. $x_a \overset{u_a}{\underset{a}{\longrightarrow}} x'_a$ *in* S_a;
  2. $x_b \overset{u_b}{\underset{b}{\longrightarrow}} x'_b$ *in* S_b;
  3. $(x_a, x_b, u_a, u_b) \in \mathcal{I}$;
- $Y_{ab} = Y_a \times Y_b$;
- $H_{ab}(x_a, x_b) = (H_a(x_a), H_b(x_b))$.

The system $S_a \times_{\mathcal{I}} S_b$ describes the concurrent evolution of systems S_a and S_b subject to the synchronization prescribed by the interconnection relation $\mathcal{I}$. The compositions represented in Figure 1.9 are but a few examples of systems that can be obtained by suitably defining $\mathcal{I}$. The interconnection relation for these examples is shown in Table 1.1. The very general notion of composition in Definition 1.6 was not chosen for the mere sake of generality. The effort placed by the reader in becoming acquainted with this notion of composition will be rewarded later in the book with simple proofs of important results.

System	$(x_a, x_b, u_a, u_b) \in \mathcal{I}$ if
S_c	$H_a(x_a) = u_b$
S_d	$H_a(x_a) = u_b$ and $H_b(x_b) = u_b$
S_e	$u_a = u_b$
S_f	$H_a(x_a) = H_b(x_b)$

Table 1.1. Description of the interconnection relation $\mathcal{I}$ for the compositions represented in Figure 1.9.

The compositions illustrated in Figure 1.9 enforce synchronization only through inputs and outputs. However, the relation $\mathcal{I}$ enables also the use of the state for synchronization. This will be essential when we discuss problems of control. The way in which transitions of $S_a \times_{\mathcal{I}} S_b$ are constructed from transitions of S_a and S_b tells us that:

$$\mathcal{B}(S_a \times_{\mathcal{I}} S_b) \subseteq \mathcal{B}(S_a) \times \mathcal{B}(S_b) \tag{1.20}$$

with equality holding for the trivial interconnection relation defined by $\mathcal{I} = X_a \times X_b \times U_a \times U_b$. In the later case, we denote $S_a \times_{\mathcal{I}} S_b$ simply by $S_a \times S_b$.

When composing systems S_a and S_b having the same set of outputs and using an interconnection relation satisfying:

$$(x_a, x_b) \in \pi_X(\mathcal{I}) \implies H_a(x_a) = H_b(x_b)$$

there is a certain redundancy in $S_a \times_{\mathcal{I}} S_b$. The output set is $Y_a \times Y_b$ with $Y_a = Y_b$ and the output at every state $(x_a, x_b) \in X_{ab}$ is $(H_a(x_a), H_b(x_b))$ with $H_a(x_a) = H_b(x_b)$. Under these circumstances we simplify $S_a \times_{\mathcal{I}} S_b$ by redefining Y_{ab} to be:

$$Y_{ab} = Y_a = Y_b$$

and by redefining H_{ab} to be:

$$H_{ab}(x_a, x_b) = H_a(x_a) = H_b(x_b).$$

Equation (1.20) now tells us that the behavior of $S_a \times_{\mathcal{I}} S_b$ is related to be behavior of S_a and S_b by:

$$\mathcal{B}(S_a \times_{\mathcal{I}} S_b) \subseteq \mathcal{B}(S_a) \qquad \mathcal{B}(S_a \times_{\mathcal{I}} S_b) \subseteq \mathcal{B}(S_b).$$

Example 1.7. We now return to the communication protocol example to illustrate system composition. If S_a is the system describing the sender (see Figure 1.3) and if S_b is the system describing the receiver (see Figure 1.4), we can construct $S_a \times_{\mathcal{I}} S_b$ by choosing $\mathcal{I}$ to be the set of all quadruples $(x_a, x_b, u_a, u_b) \in X_a \times X_b \times U_a \times U_b$ satisfying $H_a(x_a) = H_b(x_b)$ and corresponding to one of the interconnections represented in Figure 1.9. The resulting finite-state system is represented in Figure 1.10.

A possible specification for the communication protocol is that in every external behavior of $S_a \times_{\mathcal{I}} S_b$, `tx/rx` is always followed, although not necessarily immediately, by `ack`. This would require that any transmission is eventually correctly received. By inspecting Figure 1.10, we see that it is possible to cycle between the states `(send data, receive data)` and `(bad tx, bad rx)` indefinitely, thus implying that `tx/rx` may not be followed by `ack`. Unless additional assumptions are made about the communication channel, it is not possible to guarantee that transmitted messages are eventually received. In this example we could determine that $S_a \times_{\mathcal{I}} S_b$ does not conform to the specification simply by inspecting $S_a \times_{\mathcal{I}} S_b$. When dealing with large systems, analyzing $S_a \times_{\mathcal{I}} S_b$ by inspection is no longer possible and one has to resort to algorithmic verification techniques such as the ones presented in Part II. $\triangleleft$

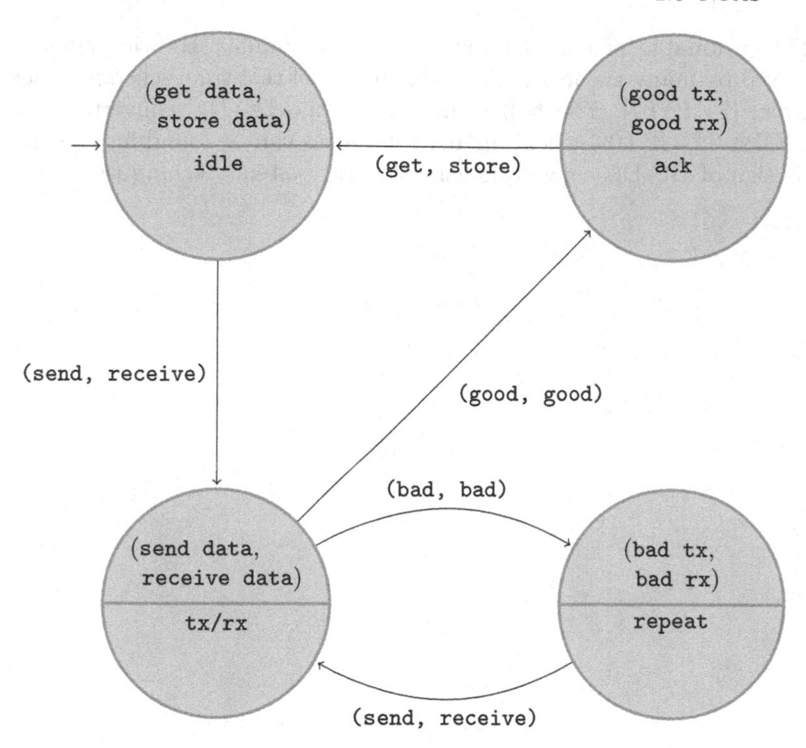

Fig. 1.10. Graphical representation of the composition of the finite-state system displayed in Figure 1.3 (modeling the sender) with the finite-state system displayed in Figure 1.4 (modeling the receiver).

1.5 Notes

The notion of system introduced in this chapter is a blend between the automata theoretic models used for the verification of software systems with the differential equation models used in control. Owing to its automata descent, the dynamics is described by a relation instead of a function. The flexibility afforded by relations will be used in Parts II, III, and IV to model the effect of disturbances through nondeterminism. Following the current practice in control theory, outputs depend on the states rather than on the inputs. This choice is motivated by the models used in control theory where the state has physical meaning which is essential to define specifications. However, as described in Section 1.2, the inputs can easily be extracted from the outputs by suitable redefining the ingredients of a system.

The separation of variables into inputs, states, and outputs is a modeling decision that is not always easy to make, especially when dealing with software. Although this practice seems to be well rooted in the computer science and control theory traditions, there exist alternative modeling formalisms that do not require such distinctions [PW97].

The national income model was proposed in [Sam39]. Hybrid systems have been used by many authors to study problems of real-time scheduling see, for example, [FKPY07]. The hybrid model for the DC-DC converter is taken from [GPM04, GPT09] where further references can be found for the analysis and design of DC-DC converters using hybrid systems techniques.

2

Verification problems

Systems are mathematical models of dynamical phenomena that allow for rigorous analysis. In this chapter we describe the two kinds of verification problems that are considered in this book.

2.1 $S_a \cong S_b$

The first verification problem is the *equivalence problem*.

Problem 2.1 (Equivalence). Given systems S_a and S_b and a notion of equivalence between systems, when is S_a *equivalent* to S_b?

If one denotes system equivalence by the symbol $\cong$, then Problem 2.1 asks when the following relationship holds:

$$S_a \cong S_b.$$

Several different analysis and verification problems arising in the design of complex systems can be casted as instances of the equivalence problem. This can be done for systems that have already been designed as well as for systems that have not yet, or have only been partially designed. In the former case, we regard S_a as a model of the system that has already been designed and S_b as a model of the specification. A positive answer to the equivalence problem would then imply that the design conforms to the specification. In the later case, we regard S_a and S_b as potential models of the same dynamical phenomenon and seek to determine if both models are equivalent. A positive answer to the equivalence problem would imply that any of the models could be used to complete the design at hand. In both cases we are implicitly assuming that one of the models is much simpler than the other. If S_b describes the specification then it is natural to expect that it should be much easier to construct S_b than S_a. When S_a and S_b are both models for the same system being designed, S_b being a much simpler model than S_a would guarantee that

P. Tabuada, *Verification and Control of Hybrid Systems: A Symbolic Approach*,
DOI: 10.1007/978-1-4419-0224-5_2, © Springer Science + Business Media, LLC 2009

the remaining design could be accomplished with greater ease by working with the simpler model S_b. This observation immediately places some restrictions on the notions of equivalence as they need to treat as equivalent, system S_a and the much simpler system S_b.

In this book we distinguish between two different kinds of equivalence: exact and approximate. While exact equivalence can be used for finite-state and infinite-state systems, approximate equivalence is more natural in the context of infinite-state systems describing dynamical, control, or hybrid systems. Exact equivalence requires the outputs of equivalent systems to be exactly the same while approximate equivalence relaxes this requirement by allowing the outputs to differ up to some specified precision. It is shown in Part IV that the additional flexibility afforded by approximate equivalence results in a larger class of infinite-state systems having equivalent finite-state symbolic models.

2.2 $S_a \preceq S_b$

In many circumstances the equivalence problem may be too demanding. If S_b is a model for the specification, it may be impossible to design a system S_a that is equivalent to S_b. However, S_a may still satisfy the specification in a weaker sense captured by the *pre-order problem.*

Problem 2.2 (Pre-order). Given systems S_a and S_b and a pre-order[1] between systems, when does S_a *precede* S_b?

If one denotes the pre-order by the symbol $\preceq$, then Problem 2.2 asks when the following relationship holds:

$$S_a \preceq S_b.$$

Intuitively, $S_a \preceq S_b$ is interpreted as S_a being "included" in S_b. The exact meaning of "included" will depend on the particular pre-order being used. As was the case with equivalence we will consider exact and approximate pre-orders, the later being a generalization of the former.

[1] Recall that a pre-order is a relation which is reflexive and transitive. See the Appendix for more details on pre-orders.

3

Control problems

Formal models and techniques can be used, not only for the verification of systems, but also for its design. In this book we consider design problems for equivalence and pre-order relations.

3.1 $S_c \times_{\mathcal{I}} S_a \cong S_b$

The *control problem for equivalence* formalizes the essence of design.

Problem 3.1 (Control for equivalence). Given systems S_a and S_b, and given a notion of equivalence between systems, when does it exist and how can we construct a system S_c and an interconnection relation $\mathcal{I}$ such that:

$$S_c \times_{\mathcal{I}} S_a \cong S_b. \tag{3.1}$$

System S_b is typically a model of the specification that is to be enforced, on a given platform modeled by S_a, through the design of S_c. The control problem for equivalence makes sense at different levels of design. System S_a could model a hardware platform and S_c could model an operating system to be designed or S_a could model the hardware platform equipped with the operating system and S_c could be the desired midleware or application level software. When system S_c is designed so as to enforce (3.1), formal verification is not necessary to prove the equivalence between the designed system $S_c \times_{\mathcal{I}} S_a$ and the specification S_b. This is one of the main advantages of the use of formal methods for the design of complex engineered systems.

In this book we present methods and techniques for the solution of Problem 3.1 using exact and approximate notions of equivalence.

P. Tabuada, *Verification and Control of Hybrid Systems: A Symbolic Approach*,
DOI: 10.1007/978-1-4419-0224-5_3, © Springer Science + Business Media, LLC 2009

3.2 $S_c \times_{\mathcal{I}} S_a \preceq S_b$

The same reasons that lead us to consider pre-orders for verification problems suggest that one should also consider the *control problem for pre-order.*

Problem 3.2 (Control for pre-order). Given systems S_a and S_b and given a pre-order between systems, when does it exist and how can we construct a system S_c and an interconnection relation $\mathcal{I}$ such that:

$$S_c \times_{\mathcal{I}} S_a \preceq S_b.$$

From a mathematical point of view one can also consider the problem of designing S_c so that $S_b \preceq S_c \times_{\mathcal{I}} S_a$ holds. However, since composing S_c with S_a results in a system that is more constrained than S_a, either $S_b \preceq S_a$ already holds, in which case we do not need to design S_c, or $S_b \preceq S_a$ does not hold and no matter which S_c we use to further constrain S_a, we will never achieve $S_b \preceq S_c \times_{\mathcal{I}} S_a$.

Part II

Finite systems

4

Exact system relationships

The verification problem for a system S_a and a model of desired behavior S_b, asks the fundamental question of whether S_a is either equivalent to the desired system ($S_a \cong S_b$) or contained in the desired system ($S_a \preceq S_b$). The answer to such questions will always depend on what we mean by system equivalence and system containment. In this chapter, we give various precise definitions for such relationships between two systems.

Notation

Every relation $Q \subseteq Z_a \times Z_b$, admits $Q^{-1} = \{(z_b, z_a) \in Z_b \times Z_a \mid (z_a, z_b) \in Q\}$ as its inverse relation. Moreover, if $R \subseteq Z_b \times Z_c$ is also a relation, we denote by $R \circ Q$ the composite relation defined by all the pairs $(z_a, z_c) \in Z_a \times Z_c$ for which there exists $z_b \in Z_b$ such that $(z_a, z_b) \in Q$ and $(z_b, z_c) \in R$. When Q is an equivalence relation on a set Z, we denote by $[z]$ the equivalence class of $z \in Z$, by Z/Q the set of all equivalence classes, and by $\pi_Q : Z \to Z/Q$ the natural projection map taking a point $z \in Z$ to its equivalence class $\pi(z) = [z] \in Z/Q$. We say that an equivalence relation is finite when it has finitely many equivalence classes.

Given a collection of sets $\mathcal{Z} = \{Z_i\}_{i \in I}$ and an element $W \in \mathcal{Z}$, we say that W is maximal (with respect to set inclusion) if for every $Z_i \in \mathcal{Z}$ we have $Z_i \subseteq W$.

The pre-image of a set $K \subseteq W$ under a map $f : Z \to W$ is denoted by $f^{-1}(K)$ and defined as the set $f^{-1}(K) = \{z \in Z \mid f(z) \in K\}$.

For a set Z, Z^* and Z^ω denote the set of all finite and infinite strings, respectively, obtained by concatenating elements in Z. An element $\mathsf{z} \in Z^*$ can thus be seen as a map $\mathsf{z} : \{0, 1, 2, \ldots, n\} \to Z$ represented by $\mathsf{z} = z_0 z_1 z_2 \ldots z_n$ with $\mathsf{z}(i) = z_i$, $i \in \{0, 1, 2, \ldots, n\}$. Similarly, an element $\mathsf{z} \in Z^\omega$ is a map $\mathsf{z} : \mathbb{N}_0 \to Z$ represented by $\mathsf{z} = z_0 z_1 z_2 \ldots$ with $\mathsf{z}(i) = z_i$, $i \in \mathbb{N}_0$.

P. Tabuada, *Verification and Control of Hybrid Systems: A Symbolic Approach*,
DOI: 10.1007/978-1-4419-0224-5_4, © Springer Science + Business Media, LLC 2009

4.1 Behavioral relationships

The first natural relationship between two systems S_a and S_b requires every external behavior of system S_a to also be an external behavior of system S_b.

Definition 4.1 (Behavioral inclusion). *Given two systems S_a and S_b with $Y_a = Y_b$, we say that S_a is* behaviorally included *in S_b, denoted by $S_a \preceq_{\mathcal{B}} S_b$, if $\mathcal{B}^\omega(S_a) \subseteq \mathcal{B}^\omega(S_b)$.*

If S_a is a model of a system that we wish to analyze, and S_b is a model of the desired behavior, then showing that $S_a \preceq_{\mathcal{B}} S_b$ effectively shows that all behaviors of S_a satisfy the desired specification.

Behavioral inclusion can be viewed as a *behavioral matching game* between the two systems. The game is played in alternating rounds between the two systems. If we want to show that $S_a \preceq_{\mathcal{B}} S_b$, then system S_a plays first and selects any behavior $\mathsf{y} \in \mathcal{B}^\omega(S_a)$. System S_b plays after system S_a and must then produce exactly the same behavior, *i.e.*, $\mathsf{y} \in \mathcal{B}^\omega(S_b)$. If system S_b can match every behavior that S_a produces, then S_b wins the game, resulting in $S_a \preceq_{\mathcal{B}} S_b$. On the other hand, if S_a produces a behavior $\mathsf{y} \in \mathcal{B}^\omega(S_a)$ that S_b cannot match, then clearly $\mathsf{y} \notin \mathcal{B}^\omega(S_b)$ and thus $S_a \not\preceq_{\mathcal{B}} S_b$.

Definition 4.2 (Behavioral equivalence). *Given two systems S_a and S_b with $Y_a = Y_b$, we say that S_a is* behaviorally equivalent *to S_b, denoted by $S_a \cong_{\mathcal{B}} S_b$, if $S_a \preceq_{\mathcal{B}} S_b$ and $S_b \preceq_{\mathcal{B}} S_a$.*

In other words, systems S_a and S_b are behaviorally equivalent if they have exactly the same set of behaviors, *i.e.*, $\mathcal{B}^\omega(S_a) = \mathcal{B}^\omega(S_b)$. In our game theoretic interpretation, we can play one game to show that $S_a \preceq_{\mathcal{B}} S_b$ and another game to show that $S_b \preceq_{\mathcal{B}} S_a$. If the first game is won by S_b, and the second game is won by S_a, then the systems are behaviorally equivalent. Alternatively, we can combine the two games into a single game, where in each round of the combined game we arbitrarily choose which system plays first, and which systems plays second. If, in every round, the system that plays second can always match the system that plays first, regardless of the choice, then the systems are behaviorally equivalent.

Example 4.3. As part of a larger effort to reduce pollution and alleviate traffic congestion, UCLA has an agreement with a local bus company to encourage members of the UCLA community to use public transportation. The agreement results in reduced fares for the riders and is implemented as follows. Upon entrance on the bus, any member of the UCLA community is encouraged to swipe his UCLA identification card on the fare machine. This action produces a distinctive sound that we call *ding*. Following the swipe, the reduced fare of a quarter should be deposited into the fare machine. A different sound, that we call *dong*, is then emitted to acknowledge the reception of the complete fare. A possible model S_a for the fare machine is shown in Figure 4.1.

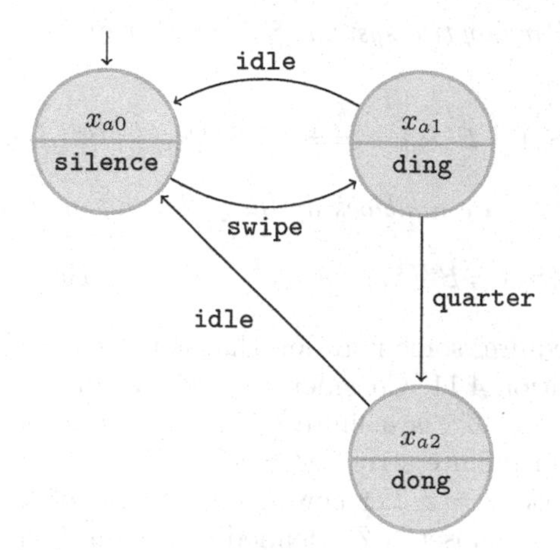

Fig. 4.1. Finite-state system modeling the bus fare machine for Example 4.3.

Figure 4.2 shows another model, S_b, that is equivalent to S_a in the sense that $S_a \cong_{\mathcal{B}} S_b$. The reader can convince himself of this fact by playing a few rounds of the behavioral matching game. $\triangleleft$

Note that given two systems, we may have $S_a \preceq_{\mathcal{B}} S_b$, $S_b \preceq_{\mathcal{B}} S_a$, $S_a \cong_{\mathcal{B}} S_b$, or neither in which case the systems are incomparable.

A sensible alternative to Definition 4.1 is to define behavioral inclusion using finite external behaviors instead of infinite external behaviors. Under a nonblocking assumption both alternatives are in fact equivalent.

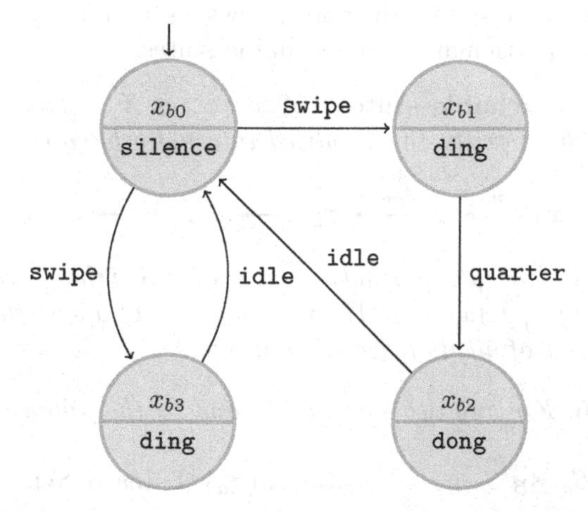

Fig. 4.2. A different finite-state system S_b modeling the bus fare machine for Example 4.3.

Proposition 4.4. *For any two systems S_a and S_b with $Y_a = Y_b$, the following implication holds:*

$$\mathcal{B}(S_a) = \mathcal{B}(S_b) \implies \mathcal{B}^\omega(S_a) = \mathcal{B}^\omega(S_b).$$

Moreover, if S_a and S_b are nonblocking the following implication also holds:

$$\mathcal{B}^\omega(S_a) = \mathcal{B}^\omega(S_b) \implies \mathcal{B}(S_a) = \mathcal{B}(S_b).$$

Proof. This proof requires some notation that is not used elsewhere except in the proof of Proposition 4.11. Consider a set Z. A finite prefix of an infinite string $\mathsf{z} = z_0 z_1 z_2 \ldots \in Z^\omega$ is a finite string $\mathsf{w} = w_0 w_1 w_2 \ldots w_k \in Z^*$ for which there exists an infinite string $\mathsf{w}' = w_0' w_1' w_2' \ldots \in Z^\omega$ such that $\mathsf{w}\mathsf{w}' = w_0 w_1 w_2 \ldots w_k w_0' w_1' w_2' \ldots = \mathsf{z}$. Let now L be a subset of Z^*. The limit of L, denoted by $\lim L$, is the subset of Z^ω defined by $\mathsf{z} \in \lim L$ iff every finite prefix of z belongs to L. With these definitions in mind we see that for any system S, $\mathcal{B}^\omega(S) = \lim \mathcal{B}(S)$. Therefore:

$$\mathcal{B}(S_a) = \mathcal{B}(S_b) \implies \lim \mathcal{B}(S_a) = \lim \mathcal{B}(S_b) \implies \mathcal{B}^\omega(S_a) = \mathcal{B}^\omega(S_b).$$

This proves the first part of the result. For the second part we need again additional notation. For a set $L \subseteq Z^\omega$ we denote by $\mathrm{Fin}\, L$, the set of all finite prefixes of strings in L. Note that we always have $\mathrm{Fin}\, \mathcal{B}^\omega(S) \subseteq \mathcal{B}(S)$ for any system S. But the opposite inclusion, $\mathcal{B}(S) \subseteq \mathrm{Fin}\, \mathcal{B}^\omega(S)$, only holds when S is nonblocking since only in this case any finite external behavior can be extended to an infinite external behavior. The second part of the result then follows from the equality $\mathrm{Fin}\, \mathcal{B}^\omega(S) = \mathcal{B}(S)$ which holds under the nonblocking assumption. □

Formally relating system behaviors allows us to formally relate system properties, and, in particular, their reachable states.

Definition 4.5 (Reachable states). *A state $x \in X$ is said to be* reachable *in a system S if there exists an initialized internal behavior:*

$$x_0 \xrightarrow{u_0} x_1 \xrightarrow{u_1} x_2 \longrightarrow \ldots \xrightarrow{u_{n-1}} x.$$

An output $y \in Y$ is said to be reachable in a system S if there exists a reachable state $x \in X$ satisfying $H(x) = y$. The reachable set of a system S, denoted by $\mathrm{Reach}(S)$*, is the set of all its reachable outputs.*

Proposition 4.6. *For any two systems S_a and S_b the following implications hold:*

$$S_a \preceq_{\mathcal{B}} S_b \implies \mathrm{Reach}(S_a) \subseteq \mathrm{Reach}(S_b),$$

$$S_a \cong_{\mathcal{B}} S_b \implies \mathrm{Reach}(S_a) = \mathrm{Reach}(S_b).$$

Proof. Let $y_a \in \text{Reach}(S_a)$. By definition of reachable output, there exists a finite external behavior $\mathsf{y} = y_0 y_1 \ldots y_n$ with $y_n = y_a$. Assuming that $S_a \preceq_{\mathcal{B}} S_b$ holds, we have $\mathsf{y} \in \mathcal{B}(S_b)$. Consequently, $y_a \in \text{Reach}(S_b)$ thus showing that $\text{Reach}(S_a) \subseteq \text{Reach}(S_b)$. The second claim follows immediately from the first since $S_a \preceq_{\mathcal{B}} S_b$ implies $\text{Reach}(S_a) \subseteq \text{Reach}(S_b)$ and $S_b \preceq_{\mathcal{B}} S_a$ implies $\text{Reach}(S_b) \subseteq \text{Reach}(S_a)$. □

The preceding proposition can be very useful in simplifying safety analysis. Given system S_a and an unsafe set $B \subset Y_a$, the *safety* problem asks whether $\text{Reach}(S_a) \cap B = \varnothing$. If the reachable set of S_a intersects any element of B, then S_a is declared unsafe, whereas if the intersection is empty, then S_a is declared safe. In the real-time scheduling example, given in Chapter 1, the unsafe set would contain only outputs corresponding to states having `error` as its finite part since such states correspond to missed deadlines. The safety problem admits an alternative formulation based on the safe outputs $G = Y_a \backslash B$. Using G, we are now interested in deciding if $\text{Reach}(S_a) \subseteq G$. In other words, the set G which encodes safe behavior is an invariant for all system behaviors.

Unfortunately, for many systems, $\text{Reach}(S_a)$ may be difficult to compute. In such cases, an alternative and simpler approach to the safety problem is to consider another system S_b with $S_a \preceq_{\mathcal{B}} S_b$ and for which $\text{Reach}(S_b)$ is easier to compute. Since $\text{Reach}(S_a) \subseteq \text{Reach}(S_b)$, if we can show that $\text{Reach}(S_b) \cap Z = \varnothing$ then, we can conclude $\text{Reach}(S_a) \cap Z = \varnothing$.

In the context of the preceding paragraph, model S_b is not being used as a model of desired behavior for S_a, but rather as a coarser model for S_a. System S_b is also known as a modeling *abstraction* of S_a, and, dually, S_a is known as a modeling *refinement* of S_b. When dealing with safety properties, it is important that abstractions are overconservative, in the sense that all behaviors of S_a are included in S_b. Note, however, that overconservative abstractions are only sufficient for safety, as the abstracted model S_b may contain behaviors that are not feasible in S_a. Therefore, if S_b is found to be unsafe then we cannot necessarily infer that S_a is also unsafe: we have no way of knowing whether the lack of safety should be attributed to behaviors of S_a or to behaviors that belong in S_b but are not feasible in S_a. If, however, $S_a \cong_{\mathcal{B}} S_b$, then clearly $\text{Reach}(S_a) = \text{Reach}(S_b)$ and therefore S_a is safe if and only if S_b is safe.

4.2 Similarity relationships

It was argued in the previous section that behavioral inclusion $S_a \preceq_{\mathcal{B}} S_b$ of two systems is important for abstracting system S_a by system S_b or for verifying that $\mathcal{B}^\omega(S_a)$ is contained in the desired behaviors $\mathcal{B}^\omega(S_b)$. However, for infinite-state systems it is difficult, from a technical point of view, to work directly with behaviors. In this section, we develop new relationships between systems which are stronger than their behavioral counterparts. The first such relationship is based on the notion of simulation relation.

Definition 4.7 (Simulation Relation). *Consider systems S_a and S_b with $Y_a = Y_b$. A relation $R \subseteq X_a \times X_b$ is a* simulation relation *from S_a to S_b if the following three conditions are satisfied:*

1. *for every $x_{a0} \in X_{a0}$, there exists $x_{b0} \in X_{b0}$ with $(x_{a0}, x_{b0}) \in R$;*
2. *for every $(x_a, x_b) \in R$ we have $H_a(x_a) = H_b(x_b)$;*
3. *for every $(x_a, x_b) \in R$ we have that:*
   $x_a \xrightarrow[a]{u_a} x'_a$ in S_a implies the existence of $x_b \xrightarrow[b]{u_b} x'_b$ in S_b satisfying $(x'_a, x'_b) \in R$.

Intuitively, a simulation relation $R \subseteq X_a \times X_b$ captures which states of S_a are *simulated* by which states of S_b. Therefore, if $(x_a, x_b) \in R$, then state x_a of system S_a is simulated by state x_b of system S_b. Note that there is a clear order between the two systems as system S_b simulates system S_a.

With this intuition, we revisit the three requirements of Definition 4.7. First, the simulation relation must respect initial states which means that every initial state of S_a is simulated by some initial state of S_b. Second, the simulation relation must respect observations which means that if x_b simulates x_a, then the output at x_b should be exactly the same as the output at x_a. The final and most challenging requirement is that the simulation relation must respect transitions. Informally speaking, the requirement states that no matter which transition $x_a \xrightarrow[a]{u_a} x'_a$ system S_a chooses, such transition can be matched by some transition $x_b \xrightarrow[b]{u_b} x'_b$ of S_b, possibly labeled by a different input. Matching means that the successor states remain in the simulation relation, hence $(x'_a, x'_b) \in R$. This matching of transitions can also be given a game theoretic interpretation in which system S_a chooses a transition $x_a \xrightarrow[a]{u_a} x'_a$ that system S_b needs to match with a transition $x_b \xrightarrow[b]{u_b} x'_b$ for which $(x'_a, x'_b) \in R$. If system S_b can match all the choices of transition made by player a, then R is a simulation relation. Note that while behavior containment required the matching of behaviors, simulation requires the more detailed matching of individual transitions.

Definition 4.8 (Simulation). *Given two systems S_a and S_b with $Y_a = Y_b$, we say that S_a is* simulated *by S_b or that S_b* simulates *S_a, denoted by $S_a \preceq_{\mathcal{S}} S_b$, if there exists a simulation relation from S_a to S_b.*

The simple observation that behaviors are no more than a sequence of transitions leads to the next result.

Proposition 4.9. *For any two systems S_a and S_b with $Y_a = Y_b$, the following implication holds:*

$$S_a \preceq_{\mathcal{S}} S_b \implies S_a \preceq_{\mathcal{B}} S_b. \tag{4.1}$$

Proof. Assume that $S_a \preceq_S S_b$ and let R be the simulation relation. Let $\mathsf{y} \in \mathcal{B}^\omega(S_a)$ and let:

$$x_{a0} \xrightarrow[a]{u_{a0}} x_{a1} \xrightarrow[a]{u_{a1}} x_{a2} \xrightarrow[a]{u_{a2}} \cdots \tag{4.2}$$

be the corresponding initialized internal behavior. We now construct an initialized internal behavior:

$$x_{b0} \xrightarrow[b]{u_{b0}} x_{b1} \xrightarrow[b]{u_{b1}} x_{b2} \xrightarrow[b]{u_{b2}} \cdots \tag{4.3}$$

of S_b satisfying $H_b(x_{bi}) = H_a(x_{ai})$ for $i \in \mathbb{N}_0$. Since (4.2) is an initialized internal behavior, $x_{a0} \in X_{a0}$ and by definition of simulation relation there exists $x_{b0} \in X_{b0}$ such that $(x_{a0}, x_{b0}) \in R$. It now follows again from the definition of simulation relation that, since $(x_{a0}, x_{b0}) \in R$ and $x_{a0} \xrightarrow[a]{u_{a0}} x_{a1}$, there must exist a transition $x_{b0} \xrightarrow[b]{u_{b0}} x_{b1}$ in S_b with $(x_{a1}, x_{b1}) \in R$. We can repeat the argument again by using $(x_{a1}, x_{b1}) \in R$ and $x_{a1} \xrightarrow[a]{u_{a1}} x_{a2}$ to conclude the existence of $x_{b1} \xrightarrow[b]{u_{b1}} x_{b2}$ with $(x_{a2}, x_{b2}) \in R$. By repeating this argument inductively we conclude the existence of the initialized internal behavior (4.3) satisfying $(x_{ai}, x_{bi}) \in R$ for $i \in \mathbb{N}_0$. Finally, invoking the definition of simulation relation, $(x_{ai}, x_{bi}) \in R$ implies $H_a(x_{ai}) = H_b(x_{bi})$ so that the external behavior associated with (4.3) is y. Therefore $\mathsf{y} \in \mathcal{B}^\omega(S_a)$ implies $\mathsf{y} \in \mathcal{B}^\omega(S_b)$ as desired. $\square$

As the next example illustrates, the implication (4.1) cannot, in general, be reversed.

Example 4.10. Consider again the systems S_a and S_b, represented in Figures 4.1 and 4.2, respectively, and modeling the bus fare machine of the previous section. Although these systems satisfy $S_a \preceq_{\mathcal{B}} S_b$ we now show that $S_a \preceq_S S_b$ fails to hold. We proceed by contradiction and try to construct a simulation relation R. Since each system only has one initial state we necessarily have $(x_{a0}, x_{b0}) \in R$. Consider now the transition $x_{a0} \xrightarrow[a]{\texttt{swipe}} x_{a1}$. There are two possible states of S_b that can be related to x_{a1}: x_{b1} and x_{b3}. The states x_{b0} and x_{b2} cannot be related to x_{a1} since they have different observations. Relating x_{b1} to x_{a1} would not result in a simulation relation since the transition $x_{a1} \xrightarrow[a]{\texttt{idle}} x_{a0}$ in S_a cannot be matched by a transition from x_{b1} in S_b. Moreover, relating x_{b3} to x_{a1} would also not result in a simulation relation since the transition $x_{a1} \xrightarrow[a]{\texttt{quarter}} x_{a2}$ in S_a cannot be matched by a transition from x_{b3} in S_b. We thus conclude that there cannot be a simulation relation from S_a to S_b. $\lhd$

Notwithstanding the negative result illustrated by the previous example, we have the following partial converse to Proposition 4.9.

Proposition 4.11. *Let S_a and S_b be systems with $Y_a = Y_b$ and assume that S_b is output deterministic. Then, the following implication holds:*

$$\mathcal{B}(S_a) \subseteq \mathcal{B}(S_b) \implies S_a \preceq_S S_b.$$

Furthermore, when S_a is nonblocking the preceding implication can be strengthened to:

$$S_a \preceq_{\mathcal{B}} S_b \implies S_a \preceq_S S_b.$$

Proof. Without loss of generality we assume that every state $x_a \in X_a$ is reachable in S_a and that every state $x_b \in X_b$ is reachable in S_b. We can make this assumption since states that are not reachable can be eliminated without changing the internal and external behavior. Define the relation $R \subseteq X_a \times X_b$ by $(x_a, x_b) \in R$ if there exists an initialized internal behavior:

$$x_{a0} \xrightarrow[a]{u_{a0}} x_{a1} \xrightarrow[a]{u_{a1}} \cdots \xrightarrow[a]{u_{ak}} x_{ak+1} = x_a \tag{4.4}$$

in S_a and an initialized internal behavior:

$$x_{b0} \xrightarrow[b]{u_{b0}} x_{b1} \xrightarrow[b]{u_{b1}} \cdots \xrightarrow[b]{u_{bk}} x_{bk+1} = x_b \tag{4.5}$$

in S_b satisfying $H_a(x_{ai}) = H_b(x_{bi})$ for $i = 0, 1, \ldots, k+1$. The rest of the proof consists in showing that R is a simulation relation from S_a to S_b.

Let $x_{a0} \in X_{a0}$ and consider the initialized internal behavior of length one given by x_{a0}. It follows from $\mathcal{B}(S_a) \subseteq \mathcal{B}(S_b)$ that $H_a(x_{a0}) = \mathsf{y} \in \mathcal{B}(S_b)$. But this implies the existence of $x_{b0} \in X_{b0}$ satisfying $H_b(x_{b0}) = \mathsf{y} = H_a(x_{a0})$. Consequently, $(x_{a0}, x_{b0}) \in R$ thus showing that the first requirement in Definition 4.7 holds.

Since the second requirement in Definition 4.7 is satisfied by construction, we only need to show the third requirement.

Let $(x_a, x_b) \in R$ and consider a transition $x_a \xrightarrow[a]{u_a} x'_a$ in S_a. This transition can be appended to the sequence of transitions (4.4) to obtain:

$$x_{a0} \xrightarrow[a]{u_{a0}} x_{a1} \xrightarrow[a]{u_{a1}} \cdots \xrightarrow[a]{u_{ak}} x_{ak+1} = x_a \xrightarrow[a]{u_a = u_{ak+1}} x'_a = x_{ak+2}.$$

The corresponding external behavior $\mathsf{y} = y_0 y_1 \ldots y_{k+2}$ with $y_i = H_a(x_{ai})$ for $i = 0, 1, \ldots, k+2$ belongs to $\mathcal{B}(S_b)$ in virtue of the assumption $\mathcal{B}(S_a) \subseteq \mathcal{B}(S_b)$. But this implies the existence of a sequence of transitions:

$$x_{b0} \xrightarrow[b]{u_{b0}} x_{b1} \xrightarrow[b]{u_{b1}} \cdots \xrightarrow[b]{u_{bk}} x_{bk+1} \xrightarrow[b]{u_{bk+1}} x_{bk+2} \tag{4.6}$$

in S_b satisfying $H_b(x_{bi}) = y_i$ for $i = 0, 1, \ldots, k+2$. It now follows from the output determinism assumption that $y_0 y_1 \ldots y_{k+1}$ determines the corresponding internal behavior in S_b uniquely and this implies that the restriction of (4.6) to its first $k+1$ transitions is given by (4.5). Therefore, we conclude the existence of the transition $x_b = x_{bk+1} \xrightarrow[b]{u_b = u_{bk+1}} x_{bk+2} = x'_b$ satisfying $(x'_a, x'_b) \in R$.

To prove the last assertion in the proposition, recall from the proof of Proposition 4.4 that $\mathrm{Fin}\, L$ is the set of all finite prefixes of strings in $L \subseteq Y^\omega$. Recall also that we always have $\mathrm{Fin}\, \mathcal{B}^\omega(S) \subseteq \mathcal{B}(S)$ but the reverse inclusion only holds when S is nonblocking. With these facts in mind, the last assertion is proved by noting that:

$$\mathcal{B}^\omega(S_a) \subseteq \mathcal{B}^\omega(S_b) \implies \mathrm{Fin}\, \mathcal{B}^\omega(S_a) \subseteq \mathrm{Fin}\, \mathcal{B}^\omega(S_b) \subseteq \mathcal{B}(S_b) \implies \mathcal{B}(S_a) \subseteq \mathcal{B}(S_b)$$

where the last implication follows from the nonblocking assumption on S_a. □

Symmetrizing the notion of simulation we arrive at bisimulation.

Definition 4.12 (Bisimulation). *Given two systems S_a and S_b with $Y_a = Y_b$, we say that S_a is* bisimilar *to S_b, denoted by $S_a \cong_S S_b$, if there exists a relation R satisfying:*

1. *R is a simulation relation from S_a to S_b;*
2. *R^{-1} is a simulation relation from S_b to S_a.*

Alternatively, we can define bisimilarity through the notion of bisimulation relation.

Definition 4.13 (Bisimulation Relation). *Consider systems S_a and S_b with $Y_a = Y_b$. A relation $R \subseteq X_a \times X_b$ is called a* bisimulation relation *between S_a and S_b if the following conditions are satisfied:*

1. *a) for every $x_{a0} \in X_{a0}$, there exists $x_{b0} \in X_{b0}$ with $(x_{a0}, x_{b0}) \in R$;*
   b) for every $x_{b0} \in X_{b0}$, there exists $x_{a0} \in X_{a0}$ with $(x_{a0}, x_{b0}) \in R$;
2. *for every $(x_a, x_b) \in R$ we have $H_a(x_a) = H_b(x_b)$;*
3. *for every $(x_a, x_b) \in R$ we have that:*
   a) $x_a \xrightarrow[a]{u_a} x'_a$ in S_a implies the existence of $x_b \xrightarrow[b]{u_b} x'_b$ in S_b satisfying $(x'_a, x'_b) \in R$;
   b) $x_b \xrightarrow[b]{u_b} x'_b$ in S_b implies the existence of $x_a \xrightarrow[a]{u_a} x'_a$ in S_a satisfying $(x'_a, x'_b) \in R$.

Systems S_a and S_b are said to be bisimilar if there exists a bisimulation relation between S_a and S_b.

The reader should convince himself that the two proposed definitions of bisimulation are in fact equivalent. The relation between behavioral equivalence and bisimulation can be obtained by combining Propositions 4.9 and 4.11.

Proposition 4.14. *For any two systems S_a and S_b with $Y_a = Y_b$, the following implication holds:*

$$S_a \cong_{\mathcal{S}} S_b \implies S_a \cong_{\mathcal{B}} S_b.$$

Moreover, if S_a and S_b are nonblocking and output deterministic then the converse is also true.

Example 4.15. The following relation shows that systems S_a and S_b in Figure 4.3 are bisimilar:

$$R = \{(x_{a0}, x_{b0}), (x_{a1}, x_{b0}), (x_{a2}, x_{b1}), (x_{a3}, x_{b2})\}.$$

Although equivalent, systems S_a and S_b have state sets of different cardinality. This flexibility offered by bisimulation will be exploited in Part III to construct finite-state systems that are bisimilar to infinite-state ones. ◁

If we have two simulation relations R and R' from a system S_a to a system S_b, it follows from the definition of simulation relation that $R \cup R'$ is still a simulation relation from S_a to S_b. Moreover, any simulation relation from S_a to S_b is necessarily contained in the relation $X_a \times X_b$. This simple observation is also true for bisimulation and implies the next result.

Proposition 4.16. *Let S_a and S_b be systems with $Y_a = Y_b$. The set of all simulation relations from S_a to S_b and the set of all bisimulation relations between S_a and S_b have maximal elements with respect to set inclusion.*

This property will be instrumental in later chapters since many questions in the context of verification and control can be reduced to questions regarding the maximal simulation or bisimulation.

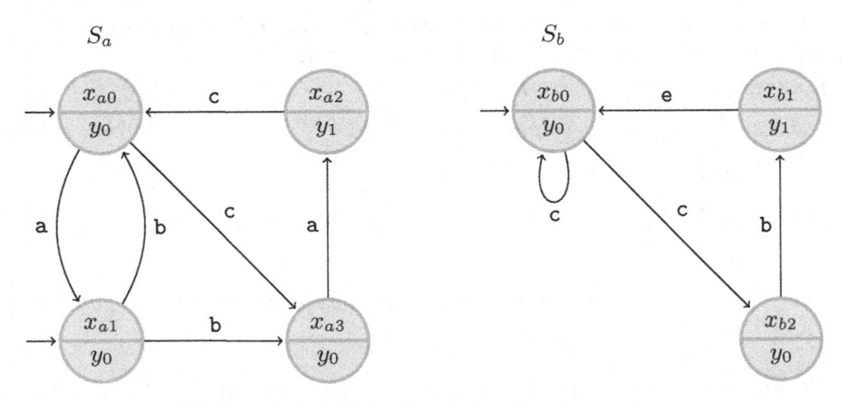

Fig. 4.3. Two bisimilar finite-state systems.

We now present a construction that will be extensively used in Chapter 7 and Chapter 8 to obtain finite-state systems from infinite-state systems. The starting point is an equivalence relation Q on the state set of a system S. If two states related by Q can be treated as equivalent, it is possible to construct a simplified description of S, a quotient system, that does not distinguish between related states.

Definition 4.17 (Quotient system). *Let $S = (X, X_0, U, \longrightarrow, Y, H)$ be a system and let Q be an equivalence relation on X such that $(x, x') \in Q$ implies $H(x) = H(x')$. The* quotient *of S by Q, denoted by $S_{/Q}$, is the system $(X_{/Q}, X_{/Q0}, U_{/Q}, \underset{/Q}{\longrightarrow}, Y_{/Q}, H_{/Q})$ consisting of:*

- $X_{/Q} = X/Q$;
- $X_{/Q0} = \{x_{/Q} \in X_{/Q} \mid x_{/Q} \cap X_0 \neq \varnothing\}$;
- $U_{/Q} = U$;
- $x_{/Q} \underset{/Q}{\overset{u}{\longrightarrow}} x'_{/Q}$ *if there exists* $x \overset{u}{\longrightarrow} x'$ *in* S *with* $x \in x_{/Q}$ *and* $x' \in x'_{/Q}$;
- $Y_{/Q} = Y$;
- $H_{/Q}(x_{/Q}) = H(x)$ *for some* $x \in x_{/Q}$.

The quotient system $S_{/Q}$ is also called a symbolic model of S since each state $x_{/Q} \in X_{/Q}$ can be regarded as a symbol representing all the states $\pi_Q^{-1}(x_{/Q}) \subseteq X$ of the original system.

System $S_{/Q}$ simulates S by construction. This is easily seen by using the graph of π_Q as the required simulation relation. This statement has a converse whenever Q is a bisimulation relation between S and S.

Theorem 4.18. *Let $S = (X, X_0, U, \longrightarrow, Y, H)$ be a system and let Q be an equivalence relation on X such that $(x, x') \in Q$ implies $H(x) = H(x')$. The relation:*

$$\Gamma(\pi_Q) = \{(x, x_{/Q}) \in X \times X_{/Q} \mid x_{/Q} = \pi_Q(x)\}$$

is a simulation relation from S to $S_{/Q}$. Moreover, $\Gamma(\pi_Q)$ is a bisimulation relation between S and $S_{/Q}$ iff Q is bisimulation relation between S and S.

Proof. The first assertion can be proved be checking that $\Gamma(\pi_Q)$ satisfies all the requirements in Definition 4.7.

We now show that when Q is a bisimulation relation between S and S, $\Gamma(\pi_Q)$ is a simulation relation from $S_{/Q}$ to S. The first two requirements in Definition 4.7 hold by construction so we focus on the third. Let $(x, x_{/Q}) \in \Gamma(\pi_Q)$, or equivalently $\pi_Q(x) = x_{/Q}$, and let $x_{/Q} \underset{/Q}{\overset{u}{\longrightarrow}} x'_{/Q}$ in $S_{/Q}$. By construction of $\underset{/Q}{\longrightarrow}$, there exists $\widehat{x} \in x_{/Q}$ and $\widehat{x}' \in x'_{/Q}$ satisfying $\widehat{x} \overset{u}{\longrightarrow} \widehat{x}'$. But then $(x, \widehat{x}) \in Q$ and since Q is a bisimulation relation, there exists $x' \in X$ satisfying $x \overset{u'}{\longrightarrow} x'$ in S and $\pi_Q(x') = \pi_Q(\widehat{x}')$. Therefore, $x \overset{u'}{\longrightarrow} x'$ is the transition we needed to conclude that $\Gamma(\pi_Q)$ satisfies the third requirement.

A similar argument shows that if $\Gamma(\pi_Q)$ is a simulation relation from $S_{/Q}$ to S then Q is a bisimulation relation. □

When the equivalence relation Q has finitely many equivalence classes, $S_{/Q}$ is guaranteed to be finite-state. Although the set of inputs $U_{/Q} = U$ will still be infinite, in general, we ask the reader to show that any finite-state system is always bisimilar to a finite-state system with a set of inputs of finite[1] cardinality.

4.3 Alternating similarity relationships

When discussing problems of control, a different kind of similarity relationship is needed. Simulation relations require the matching of transitions while in problems of control we require the existence of inputs enforcing desired transitions. We thus need a similarity relationship that captures the effect that different choices of inputs have on transitions.

Definition 4.19 (Alternating simulation relation). *Let S_a and S_b be systems with $Y_a = Y_b$. A relation $R \subseteq X_a \times X_b$ is an* alternating simulation relation *from S_a to S_b if the following three conditions are satisfied:*

1. *for every $x_{a0} \in X_{a0}$ there exists $x_{b0} \in X_{b0}$ with $(x_{a0}, x_{b0}) \in R$;*
2. *for every $(x_a, x_b) \in R$ we have $H_a(x_a) = H_b(x_b)$;*
3. *for every $(x_a, x_b) \in R$ and for every $u_a \in U_a(x_a)$ there exists $u_b \in U_b(x_b)$ such that for every $x'_b \in \mathrm{Post}_{u_b}(x_b)$ there exists $x'_a \in \mathrm{Post}_{u_a}(x_a)$ satisfying $(x'_a, x'_b) \in R$.*

As in the non-alternating case, alternating simulation relations underlie the notion of alternating simulation.

Definition 4.20 (Alternating simulation). *Given two systems S_a and S_b with $Y_a = Y_b$, we say that S_a is* alternatingly simulated *by S_b or that S_b* alternatingly simulates *S_a, denoted by $S_a \preceq_{\mathcal{AS}} S_b$, if there exists an alternating simulation relation from S_a to S_b.*

The difference between simulation relations and alternating simulation relations is best explained through an example.

Example 4.21. Consider systems S_a and S_b in Figure 4.4. The relation:

$$R = \{(x_{a0}, x_{b0}), (x_{a1}, x_{b1}), (x_{a2}, x_{b2})\}$$

is a simulation relation from S_a to S_b but not an alternating simulation. This can be seen by noting that $x_{b3} \in \mathrm{Post}_{\mathsf{a}}(x_{b0})$ in S_b although neither x_{a1} nor

[1] If nondeterminism is not a concern, we can even take the set of inputs to be a singleton and label every transition by the same input.

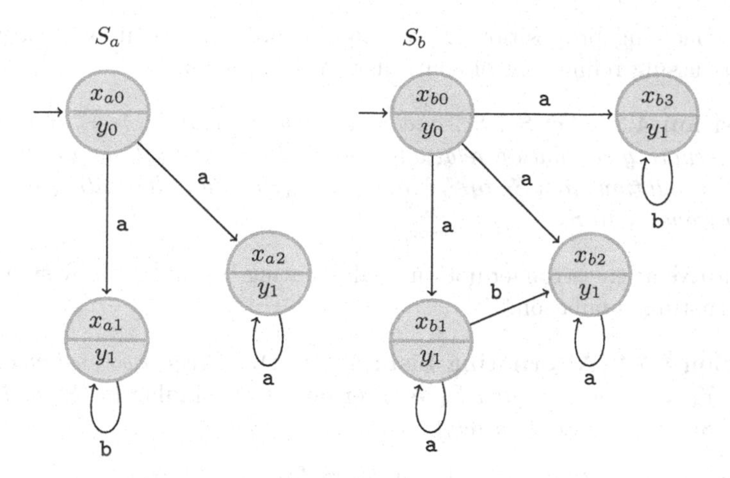

Fig. 4.4. Systems S_a and S_b for Example 4.21.

x_{a2}, the only elements in $\mathrm{Post_a}(x_{a0})$ in S_a, are related by R to x_{b3}. Conversely, the reader is invited to verify that the relation:

$$R' = \{(x_{a0}, x_{b0}), (x_{a1}, x_{b1}), (x_{a1}, x_{b2}), (x_{a1}, x_{b3})\}$$

is an alternating simulation relation from S_a to S_b. However, R' is not a simulation relation from S_a to S_b since the transition $x_{a0} \xrightarrow[a]{\mathtt{a}} x_{a2}$ in S_a cannot be matched by S_b. $\lhd$

Although alternating simulation is substantially different from simulation, these two notions coincide in the very special case of deterministic systems. This can be easily seen by noting that determinism implies $|\mathrm{Post}_{u_b}(x_b)| \leq 1$ and $|\mathrm{Post}_{u_a}(x_a)| \leq 1$ for every $u_b \in U_b(x_b)$ and $u_a \in U_a(x_a)$. Hence, the third requirement in Definition 4.19 becomes the third requirement in Definition 4.7.

When addressing problems of control in Chapter 6 we will need an alternating simulation relation relating, not only states, but also inputs.

Definition 4.22 (Extended alternating simulation relation). *Let R be an alternating simulation relation from system S_a to system S_b. The* extended alternating simulation relation $R^e \subseteq X_a \times X_b \times U_a \times U_b$ *associated with R is defined by all the quadruples $(x_a, x_b, u_a, u_b) \in X_a \times X_b \times U_a \times U_b$ for which the following three conditions hold:*

1. *$(x_a, x_b) \in R$;*
2. *$u_a \in U_a(x_a)$;*
3. *$u_b \in U_b(x_b)$ and for every $x'_b \in \mathrm{Post}_{u_b}(x_b)$ there exists $x'_a \in \mathrm{Post}_{u_a}(x_a)$ satisfying $(x'_a, x'_b) \in R$.*

Note that the third requirement in Definition 4.22 is no more than the third requirement in Definition 4.19.

The following proposition, whose simple proof we omit, will be needed when discussing refinement of controllers in Chapter 6.

Proposition 4.23. *Let S_a, S_b, and S_c be systems with $Y_a = Y_b = Y_c$. If ${}_aR_b$ is an alternating simulation relation from S_a to S_b and ${}_bR_c$ is an alternating simulation relation from S_b to S_c, then ${}_bR_c \circ {}_aR_b$ is an alternating simulation relation from S_a to S_c.*

We arrive at the stronger notion of alternating bisimulation by symmetrizing alternating simulation.

Definition 4.24 (Alternating bisimulation). *Given two systems S_a and S_b with $Y_a = Y_b$, we say that S_a is* alternatingly bisimilar *to S_b, denoted by $S_a \cong_{\mathcal{AS}} S_b$, if there exists a relation R satisfying:*

1. *R is an alternating simulation relation from S_a to S_b;*
2. *R^{-1} is an alternating simulation relation from S_b to S_a.*

The reader is asked to patiently wait until Chapter 8, where we discuss problems of controller design and refinement, to fully appreciate the role of alternating simulations and bisimulations.

4.4 Notes

The notions of simulation and bisimulation, considered in this book, closely mirror the classical homonym notions introduced by Park [Par81] and Milner [Mil89]. The main difference lies in a shift of emphasis from inputs to outputs: related states need to have the same outputs but matching transitions do not need to be labeled by the same inputs. There is an important topic in the study of simulation relations that we did not discuss: the relation between simulation and composition. Simulation relations are known to commute with composition, *i.e.*, if $S_a \preceq_{\mathcal{S}} S_b$ and $S_c \preceq_{\mathcal{S}} S_d$, then $S_a \times_{\mathcal{I}} S_c \preceq_{\mathcal{S}} S_b \times_{\mathcal{J}} S_d$ where $\mathcal{J}$ is an interconnection relation that can be computed from $\mathcal{I}$ and from the simulation relations from S_a to S_b and from S_c to S_d. In Chapters 8 and 11 we touch upon these results in the context of alternating simulation relations and controller refinement.

The notion of alternating simulation and bisimulation relation is adapted from the work of Alur and co-workers [AHKV98].

5

Verification

Verification problems can be formulated using behavioral and similarity relationships. We discuss both versions in this chapter and show how to convert verification problems using behavioral relationships into verification problems using similarity relationships. The later kind can be solved by computing fixed-points of conveniently defined operators.

Notation

For a set Z, $|Z|$ denotes the cardinality of Z and Z^ω denotes the set of all infinite strings obtained by concatenating elements in Z. An element $\mathsf{z} \in Z^\omega$ can thus be seen as a map $\mathsf{z} : \mathbb{N}_0 \to Z$ represented by $\mathsf{z} = z_0 z_1 z_2 \ldots$ with $\mathsf{z}(i) = z_i$, $i \in \mathbb{N}_0$.

Given a map $f : Z \to W$ and a subset $K \subseteq W$ we denote by $f^{-1}(K)$ the subset of Z defined by $f^{-1}(K) = \{z \in Z \mid f(z) \in K\}$. When $K \subseteq Z$, we denote by $f(K)$ the set $f(K) = \{w \in W \mid w = f(z) \text{ for some } z \in Z\}$. Also used is the map $\pi_a : X_a \times X_b \to X_a$ taking the pair (x_a, x_b) to $\pi_a(x_a, x_b) = x_a \in X_a$. The set of all subsets of Z, also known as the power set of Z, is denoted by 2^Z.

5.1 Behavioral relations

Showing that a system S_a satisfies a desired property P is a problem that recurrently appears in the analysis and design of complex engineering systems. Using the notion of system introduced in Chapter 1, we can formalize the concept of *property* as a subset of Y_a^ω. Intuitively, a property P of a system S_a allows us to separate Y_a^ω into the set of strings that satisfy P and the set of strings that do not satisfy P. Therefore, we can identify P with the set of strings that satisfies this property and thus think of a property as being a subset of Y_a^ω. This conceptually appealing formulation is difficult to be directly

P. Tabuada, *Verification and Control of Hybrid Systems: A Symbolic Approach*,
DOI: 10.1007/978-1-4419-0224-5_5,

used since writing down explicitly the set of all strings satisfying a given property P is very difficult if not impossible. Fortunately, there exist alternative and simpler ways to specify properties such as temporal logics and other formal specification formalisms. A detailed exposition of these techniques would force us to digress into the realm of temporal logics and automata theory for which good expositions are already available in the literature. Instead, we consider only properties that can be specified as $\mathcal{B}^\omega(S_b)$ for some finite-state system S_b. In order to determine if a system S_a satisfies property P we equivalently test if $\mathcal{B}^\omega(S_a)$ belongs to the set of all the infinite strings satisfying P, which is simply $\mathcal{B}^\omega(S_b)$. Hence, this formulation leads to the preorder problem $S_a \preceq_{\mathcal{B}} S_b$, formulated in Part I, that asks if the following inclusion is satisfied:

$$\mathcal{B}^\omega(S_a) \subseteq \mathcal{B}^\omega(S_b) \tag{5.1}$$

for a system S_a and a specification system S_b. From Proposition 4.11 we know that inclusion (5.1) is equivalent to:

$$S_a \preceq_{\mathcal{S}} S_b$$

whenever S_a is nonblocking and S_b is output deterministic. These are reasonable assumptions that we shall adopt. We can easily check if S_a is blocking by searching for states that have no outgoing transitions. If one such state is found, then S_a needs to be redesigned since S_a is a model for a reactive system, such as an embedded controller, that must operate uninterruptedly for arbitrarily long periods of time. Regarding the output determinism assumption on S_b, we now show that we can always construct an output deterministic system S_c that is behaviorally equivalent to S_b. The proof of this result is based on the Myhill-Nerode equivalence classes which have long been used to solve realization problems in automata theory as well as in systems and control theory.

Proposition 5.1. *For any system S_b there exists an output deterministic system S_c satisfying $S_b \cong_{\mathcal{B}} S_c$.*

Proof. In view of Proposition 4.4 it is sufficient to show the existence of an output deterministic system S_c satisfying $\mathcal{B}(S_c) = \mathcal{B}(S_b)$.

Let R be the equivalence relation on $\mathcal{B}(S_b)$ rendering $\mathsf{y} = y_0 y_1 \ldots y_k$ equivalent to $\mathsf{y}' = y'_0 y'_1 \ldots y'_l$ if $y_k = y'_l$ and for every $y \in Y_b$ we have $\mathsf{y}y = y_0 y_1 \ldots y_k y \in \mathcal{B}(S_b) \Leftrightarrow \mathsf{y}'y = y'_0 y'_1 \ldots y'_l y \in \mathcal{B}(S_b)$. Note that R has the property:

$$(\mathsf{y}, \mathsf{y}') \in R \implies (\mathsf{y}y, \mathsf{y}'y) \in R \quad \forall y \in Y_b. \tag{5.2}$$

We now define S_c starting with the set of states X_c given by $X_c = \mathcal{B}(S_b)/R$, the set of all equivalence classes induced by R. The set of initial states X_{c0} is the set of equivalence classes containing the behaviors y of unit length given by $\mathsf{y} = H_b(x_{b0})$ for every $x_{b0} \in X_{b0}$. The set of inputs is given by $U_c = Y_b$ and the transition relation is defined by $x_c \xrightarrow[c]{y} x'_c$ if x_c is the equivalence class containing a behavior y and x'_c is the equivalence class containing the behavior yy. The transition relation is well defined in view of (5.2). The set of outputs is $Y_c = Y_b$ and $H_c(x_c) = y$ when x_c is the equivalence class containing the behavior yy.

We claim that S_c is output deterministic. First, note that the cardinality of X_{c0} is the cardinality of $H_b(X_{b0})$ and every element in X_{c0} is mapped by H_c to a different output. Therefore $H_c|_{X_{c0}}$ is injective. Moreover, if $x_c \xrightarrow[c]{u_c} x'_c$, $x_c \xrightarrow[c]{u'_c} x''_c$ in S_c and $H_c(x'_c) = y = H_c(x''_c)$, it follows by definition of $\xrightarrow[c]{}$ that $u_c = y = u'_c$, x'_c is the equivalence class containing yy, x''_c is the equivalence class containing yy, and x_c is the equivalence class containing y. Therefore $x'_c = x''_c$, as desired, and the claim is proved.

Finally, $\mathcal{B}(S_c) = \mathcal{B}(S_b)$ follows by construction of S_c. □

When system S_b is finite-state, system S_c is also guaranteed to be finite-state. Therefore, there is no loss of generality[1] in assuming that the specification system is output deterministic. The behavioral inclusion problem can thus be reduced to the verification problem in the similarity context which is discussed in the next section. Similarly, the problem of verifying behavioral equivalence of two systems can be transformed into the problem of verifying bisimilarity between two systems.

5.2 Similarity relations

Existence of simulation or bisimulation relations between two finite-state systems can be studied through the fixed-points of certain operators. The perspective offered by fixed-points of operators is advantageous on two counts. Operators are a convenient mathematical abstraction allowing us to study correctness and termination of algorithms without being distracted by the implementation details. The second advantage is the possibility of implementing symbolically the algorithms defined by fixed-points. This means that explicit enumeration of the states is avoided by manipulating instead succinct representations for sets of states.

5.2.1 Simulation relations as fixed-points

Given a system S_a and a system specification S_b we are interested in determining if $S_a \preceq_S S_b$. Ideally, an affirmative answer would also provide a simulation relation from S_a to S_b. We can thus reinterpret the simulation preorder problem as the search for a simulation relation.

[1] It should be noted, however, that the cardinality of X_c can be exponential in the cardinality of X_b.

Consider the operator:

$$F : 2^{X_a \times X_b} \to 2^{X_a \times X_b}$$

defined by $(x_a, x_b) \in F(W)$, for some $W \subseteq X_a \times X_b$, if the following three conditions are satisfied:

1. $H_a(x_a) = H_b(x_b)$;
2. $(x_a, x_b) \in W$;
3. for every transition $x_a \xrightarrow[a]{u_a} x'_a$ in S_a there exists a transition $x_b \xrightarrow[b]{u_b} x'_b$ in S_b with $(x'_a, x'_b) \in W$.

The definition of F closely mirrors the definition of simulation relation. This observation justifies the second assertion of the following proposition.

Proposition 5.2. *The operator* $F : 2^{X_a \times X_b} \to 2^{X_a \times X_b}$ *satisfies:*

1. $Z \subseteq Z'$ *implies* $F(Z) \subseteq F(Z')$*;*
2. $R \subseteq X_a \times X_b$ *is a simulation relation from* S_a *to* S_b *iff* $R \subseteq F(R)$ *and* $X_{a0} \subseteq \pi_a(R \cap (X_{a0} \times X_{b0}))$*.*

Proof. If $(x_a, x_b) \in F(Z)$, we have: $H_a(x_a) = H_b(x_b)$, $(x_a, x_b) \in Z$, and $x_a \xrightarrow[a]{u_a} x'_a$ implies $x_b \xrightarrow[b]{u_b} x'_b$ with $(x'_a, x'_b) \in Z$. Let now $Z \subseteq Z'$ and note that $H_a(x_a) = H_b(x_b)$, $(x_a, x_b) \in Z'$, and $x_a \xrightarrow[a]{u_a} x'_a$ implies $x_b \xrightarrow[b]{u_b} x'_b$ with $(x'_a, x'_b) \in Z'$. Hence, $(x_a, x_b) \in F(Z')$ and we conclude that $Z \subseteq Z'$ implies $F(Z) \subseteq F(Z')$.

We now prove the second assertion. Assume that R is a simulation relation from S_a to S_b and let $(x_a, x_b) \in R$. The first requirement in the definition of simulation implies that $X_{a0} \subseteq \pi_a(R \cap (X_{a0} \times X_{b0}))$. The second requirement implies $H_a(x_a) = H_b(x_b)$. Finally, the third requirement implies that for every $x_a \xrightarrow[a]{u_a} x'_a$ we have $x_b \xrightarrow[b]{u_b} x'_b$ with $(x'_a, x'_b) \in R$. Therefore, $(x_a, x_b) \in F(R)$ which shows that $R \subseteq F(R)$.

Conversely, let R be a relation for which the inclusions $R \subseteq F(R)$ and $X_{a0} \subseteq \pi_a(R \cap (X_{a0} \times X_{b0}))$ are satisfied. From $X_{a0} \subseteq \pi_a(R \cap (X_{a0} \times X_{b0}))$ follows directly the first requirement in the definition of simulation relation. Let now $(x_a, x_b) \in R$. Then, the first requirement in the definition of F implies the second requirement in the definition of simulation relation. Finally, since for every $x_a \xrightarrow[a]{u_a} x'_a$ we have $x_b \xrightarrow[b]{u_b} x'_b$ with $(x'_a, x'_b) \in R$, the third requirement in the definition of simulation relation is also satisfied and the result is proved. □

It follows from the general results on lattice theory reviewed in the Appendix that the maximal relation R satisfying $R \subseteq F(R)$ is a fixed-point of F, *i.e.*, $F(R) = R$. This can be seen by noting that the first assertion in Proposition 5.2 and $R \subseteq F(R)$ imply $F(R) \subseteq F\big(F(R)\big)$. But since R is the maximal relation satisfying the inclusion $R \subseteq F(R)$ we necessarily have $F(R) \subseteq R$, thus concluding $F(R) = R$. Moreover, it also follows from Corollary 12.6 in the Appendix that the first assertion in Proposition 5.2 implies the existence of the maximal fixed-point of F that can be obtained by iterating F. We summarize this discussion in the following result.

Theorem 5.3. *Let S_a and S_b be systems with $Y_a = Y_b$. The maximal simulation relation from S_a to S_b is the maximal fixed-point Z, of the operator F, that satisfies $X_{a0} \subseteq \pi_a(Z \cap (X_{a0} \times X_{b0}))$. Moreover:*

$$Z = \lim_{i \to \infty} F^i(X_a \times X_b) \tag{5.3}$$

and $S_a \preceq_S S_b$ iff $X_{a0} \subseteq \pi_a(Z \cap (X_{a0} \times X_{b0}))$.

For finite-state systems, the limit (5.3) is reached after finitely many iterations and Z can be computed in time polynomial in $|X_a||X_b|$.

Example 5.4. Consider system S_a and system S_b represented in Figure 4.3. The states in $X_a \times X_b$ are displayed in Figure 5.2. To simplify the presentation we show in Figure 5.3 the result of iterating F over $X_a \times X_b$ without labeling the states. A fixed-point is reached after two iterations and the resulting simulation relation is the one presented in Example 4.15. ◁

Example 5.5. Consider again the bus fare machine system in Figure 4.1, that we denote by S_a, and suppose that we want to verify $S_a \preceq_B S_b$ with respect to the specification S_b given in Figure 5.1. This specification allows the bus riders to swipe their card and trigger the ding sound but prevents them from dropping a quarter and hearing the dong sound.

System S_a is non-blocking and S_b is output deterministic, hence $S_a \preceq_B S_b$ is equivalent to $S_a \preceq_S S_b$. To test $S_a \preceq_S S_b$ we iterate F over $X_a \times X_b$. The states in $X_a \times X_b$ are displayed in Figure 5.4. To simplify the presentation we show in Figure 5.5 the result of iterating F over $X_a \times X_b$ without labeling the states. After three iterations we obtain $F^3(X_a \times X_b) = \varnothing$ thus concluding $S_a \not\preceq_S S_b$ and consequently $S_a \not\preceq_B S_b$. ◁

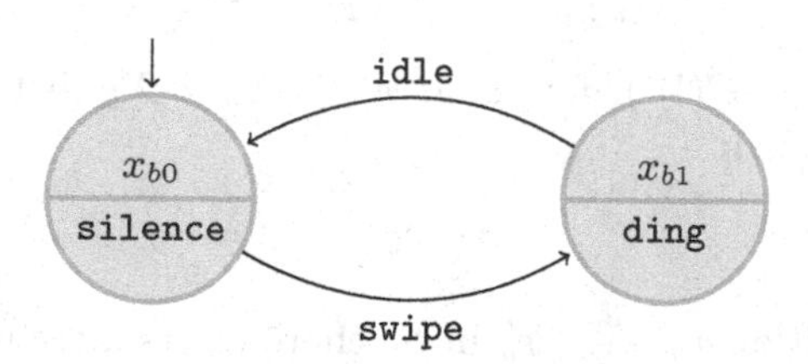

Fig. 5.1. Finite-state system S_b describing the specification for Example 5.5.

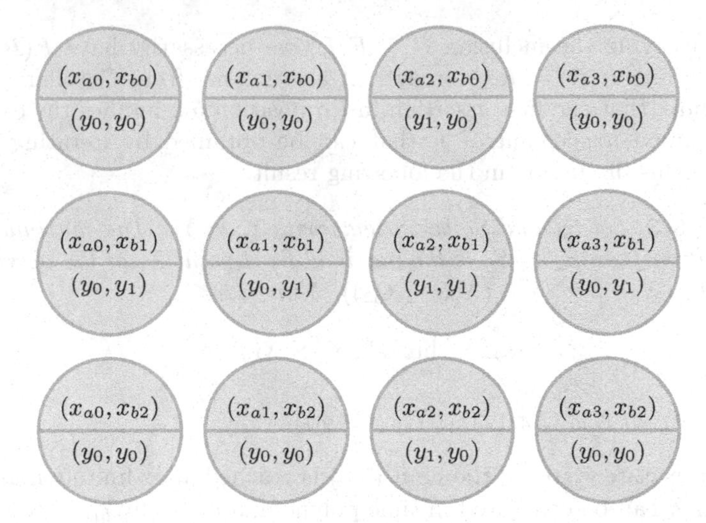

Fig. 5.2. States in the set $X_a \times X_b$ over which the operator F acts.

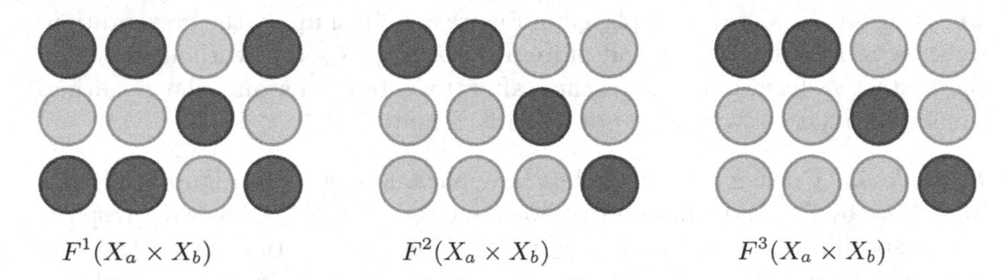

Fig. 5.3. Computation of the maximal simulation relation from S_a to S_b. The image of the operator F is represented by the dark-colored states. For conciseness, the states are not labeled and the corresponding labels can be found in Figure 5.2.

5.2.2 Bisimulation relations as fixed-points

Verifying bisimilarity between two systems can be solved by strengthening the operator F to the operator:

$$G : 2^{X_a \times X_b} \to 2^{X_a \times X_b}$$

defined by $(x_a, x_b) \in G(W)$, for some $W \subseteq X_a \times X_b$, if the following four conditions are satisfied:

1. $H_a(x_a) = H_b(x_b)$;
2. $(x_a, x_b) \in W$;
3. for every transition $x_a \xrightarrow[a]{u_a} x'_a$ in S_a there exists a transition $x_b \xrightarrow[b]{u_b} x'_b$ in S_b with $(x'_a, x'_b) \in W$;
4. for every transition $x_b \xrightarrow[b]{u_b} x'_b$ in S_b there exists a transition $x_a \xrightarrow[a]{u_a} x'_a$ in S_a with $(x'_a, x'_b) \in W$.

Reasoning in the same manner as in the simulation case, we can characterize the maximal bisimulation relation between two systems as the maximal fixed-point of G.

Theorem 5.6. *Let S_a and S_b be systems with $Y_a = Y_b$. The maximal bisimulation relation between S_a and S_b is the maximal fixed-point Z, of the operator G, that satisfies $X_{a0} \subseteq \pi_a(Z \cap (X_{a0} \times X_{b0}))$ and $X_{b0} \subseteq \pi_b(Z \cap (X_{a0} \times X_{b0}))$. Moreover:*

$$Z = \lim_{i \to \infty} G^i(X_a \times X_b) \tag{5.4}$$

and $S_a \cong_S S_b$ iff $X_{a0} \subseteq \pi_a(Z \cap (X_{a0} \times X_{b0}))$ and $X_{b0} \subseteq \pi_b(Z \cap (X_{a0} \times X_{b0}))$.

The iteration of G is guaranteed to converge in time polynomial in $|X_a||X_b|$ when S_a and S_b are finite-state systems. For infinite-state systems, the limit (5.4) may not be reached in finitely many steps so that (5.4) only defines a semi-algorithm. In Part III we will use a variation of this algorithm to compute the maximal bisimulation relation between an infinite-state system S and itself as an intermediate step in the construction of a finite-state abstraction bisimilar to S.

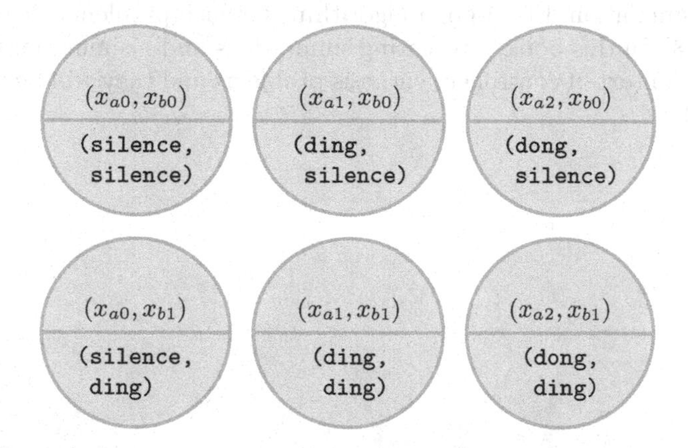

Fig. 5.4. States in the set $X_a \times X_b$ over which the operator F acts.

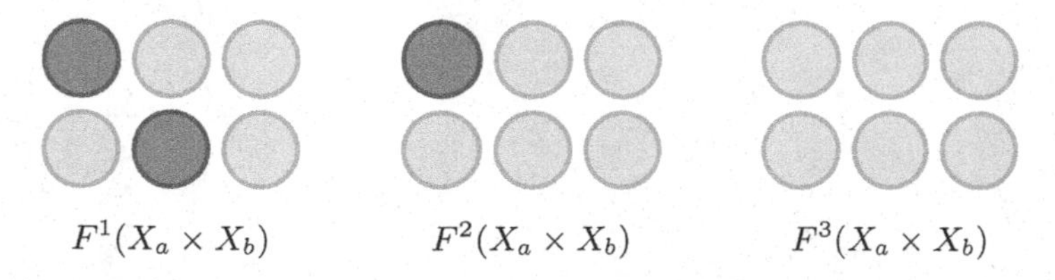

Fig. 5.5. Computation of the maximal simulation relation from S_a to S_b. The image of the operator F is represented by the dark-colored states. For conciseness, the states are not labeled and the corresponding labels can be found in Figure 5.4.

5.3 Notes

In this chapter we used systems to model both the system to be verified and the specification. In practice, however, formal specifications mechanisms such as temporal logics are preferred to systems since they are much more succinct. Modeling specifications as systems allowed us to give a flavor of the existing automated verification techniques in a concise and self-contained manner. In particular, we avoided discussing liveness specifications and its associated automata theoretic constructions. Liveness, however, will surface in Chapter 6 when discussing the synthesis of controllers enforcing reachability specifications. Chapter 6, devoted to controller synthesis problems, can be seen as a complement to this chapter since controller synthesis algorithms also answer verification questions: when a property is satisfied, no controller is necessary! The interested reader is referred to [CGP99], and [BBF$^+$01] as entry points into the rich world of formal verification.

The proof of Proposition 5.1 is based on Myhill-Nerode's equivalence classes introduced in [Ner58].

Although we discussed verification problems for simulation and bisimulation, one could also conceive verification problems using alternating simulation and bisimulation. Fixed-point algorithms for such problems are described in [AHKV98]. In this book, alternating simulations and bisimulations will appear in the context of controller synthesis problems and they will be explicitly constructed.

6

Control

Whenever a system S_a fails to conform to its specification S_b, in the sense that $S_a \not\preceq S_b$, we may ask if there exists another system S_c, the controller, such that $S_c \times_{\mathcal{I}} S_a \preceq S_b$ or even $S_c \times_{\mathcal{I}} S_a \cong S_b$. In this chapter we discuss these control problems in the behavioral and similarity contexts. We show how to reduce controller synthesis problems from the behavioral context to the similarity context and we solve the later by computing fixed-points of suitably defined operators. In addition to these general control problems we also present fixed-point solutions specialized for safety and reachability control problems that frequently arise in applications.

Notation

For a set Z, Z^* and Z^ω denote the set of all finite and infinite strings, respectively, obtained by concatenating elements in Z. An element $\mathsf{z} \in Z^*$ can thus be seen as a map $\mathsf{z} : \{0,1,2,\ldots,n\} \to Z$ represented by $\mathsf{z} = z_0 z_1 z_2 \ldots z_n$ with $\mathsf{z}(i) = z_i$, $i \in \{0,1,2,\ldots,n\}$. Similarly, an element $\mathsf{z} \in Z^\omega$ is a map $\mathsf{z} : \mathbb{N}_0 \to Z$ represented by $\mathsf{z} = z_0 z_1 z_2 \ldots$ with $\mathsf{z}(i) = z_i$, $i \in \mathbb{N}_0$. A string $\mathsf{z} \in L \subseteq Z^* \cup Z^\omega$ is said to be maximal if $\mathsf{z} \in Z^\omega$ or if $\mathsf{z} = z_0 z_1 \ldots z_k \in Z^*$ and there exists no string $\mathsf{w} = w_0 w_1 \ldots w_k w_{k+1} \in L$ satisfying $z_i = w_i$ for $i = 0, 1, \ldots, k$.

The natural projection taking $(x_a, x_b) \in X_a \times X_b$ to $x_a \in X_a$ is denoted by $\pi_a : X_a \times X_b \to X_a$. Similarly, $\pi_b : X_a \times X_b \to X_b$ denotes the natural projection taking $(x_a, x_b) \in X_a \times X_b$ to $x_b \in X_b$. The map $\pi_X : X_a \times X_b \times U_a \times U_b \to X_a \times X_b$ is also a projection and sends the quadruple $(x_a, x_b, u_a, u_b) \in X_a \times X_b \times U_a \times U_b$ to the pair $(x_a, x_b) \in X_a \times X_b$. The set of all subsets of Z, also known as the power set of Z, is denoted by 2^Z.

P. Tabuada, *Verification and Control of Hybrid Systems: A Symbolic Approach*,
DOI: 10.1007/978-1-4419-0224-5_6, © Springer Science + Business Media, LLC 2009

6.1 Feedback composition

The notion of system introduced in Part I made no claims regarding the semantics of the set U of inputs. While for some systems, the elements of U that are fed into a system can be suitably chosen, for other systems this choice is not possible. In the literature, two different approaches to the modeling of U coexist. The set of inputs U can be treated as the disjoint union of U_c and U_d, *i.e.*, $U = U_c \uplus U_d$ with U_c modeling the inputs under the designer's control (controllable) and U_d modeling the inputs beyond the designer's control (uncontrollable). Under this paradigm the effect of controllable and uncontrollable inputs is interleaved or turn-based since a transition $x \xrightarrow{u} x'$ will either be labeled by a controllable or by an uncontrollable input u. The other approach consists in describing U as the product $U = C \times D$ with C modeling the control inputs and D modeling the disturbance or adversarial inputs. In this case, starting from a state x and choosing a control input $c \in C$ leads to a transition $x \xrightarrow{c,d} x'$ in which the reached state x' depends on the choice of disturbance input $d \in D$ which is unknown and thus assumed adversarial. In this paradigm, the effect of the control and disturbance is concurrent instead of being interleaved or turn-based. We follow the concurrent approach since this is the natural paradigm for continuous-time control systems and it will be inherited by its finite-state models discussed in Parts III and IV. We do not model disturbance inputs explicitly but rather implicitly through the nondeterminism of the transition relation. This means that the disturbance has the power to decide which c-successor of a state x is reached when a control input c is chosen at the state x.

The notion of controller can be formalized in several different ways. We could regard a controller as a mechanism that determines which input should be fed into the system being controlled based on observed states[1]. This intuitive description has one important limitation: there may be more than one input that leads to a correct or desirable behavior. We thus revise the concept of controller to a mechanism that determines which inputs can be fed to the controlled system based on a sequence of observed outputs. Mathematically, this can be described by a map:

$$\phi : X^* \to 2^U$$

transforming sequences of outputs into sets of inputs. A sequence of transitions:

$$x_0 \xrightarrow{u_0} x_1 \xrightarrow{u_1} x_2 \xrightarrow{u_2} \ldots \xrightarrow{u_{n-1}} x_n$$

would then be an internal behavior of the controlled system provided that $u_k \in \phi(x_0 x_1 \ldots x_k)$ for every $k \in \{0, 1, \ldots, n-1\}$. Although this notion of controller is conceptually very pleasing, for operational reasons we restrict attention, in this chapter, to controllers $\phi : X^* \to 2^U$ that can be described

[1] To simplify the discussion, we assume $Y = X$ and $H = 1_X$.

by a finite-state system S_c. To understand how this can be done we need to digress into feedback composition.

If the effect of applying $\phi : X_a^* \to 2^{U_a}$ to a system S_a is to be described by $S_c \times_{\mathcal{I}} S_a$, we need to elaborate on the kind of interconnection relation that is appropriate for control. Among the several different possibilities we shall require $\mathcal{I}$ to be the extended relation R^e of an alternating simulation relation R from S_c to S_a. This choice renders the results that follow conceptually simple.

Definition 6.1 (Feedback composition). *A system S_c is said to be* feedback composable *with a system S_a if there exists an alternating simulation relation R from S_c to S_a. When S_c is feedback composable with S_a, the feedback composition of S_c and S_a, with interconnection relation $\mathcal{F} = R^e$, is given by $S_c \times_{\mathcal{F}} S_a$.*

The term feedback is justified by the following interpretation of $S_c \times_{\mathcal{F}} S_a$. Assume that $S_c \times_{\mathcal{F}} S_a$ is at the state $(x_c, x_a) \in R$. Controller S_c offers to execute any of the inputs $u_c \in U_c(x_c)$. System S_a responds by selecting any input $u_a \in U_a(x_a)$ satisfying $(x_c, x_a, u_c, u_a) \in \mathcal{F}$ and by taking any transition $x_a \xrightarrow[a]{u_a} x_a'$ labeled by the chosen input u_a. This transition then triggers a matching transition by the controller. This means that S_c measures the new state x_a' of S_a and takes a transition $x_c \xrightarrow[c]{u_c} x_c'$ satisfying $(x_a', x_c') \in R$. Existence of the matching transition is guaranteed by the fact that R is an alternating simulation relation. We can thus interpret an internal behavior of $S_c \times_{\mathcal{F}} S_a$ as being the result of a feedback process during which the controller offers a set of inputs, measures the state of S_a, updates its own state, offers again a new set of inputs based on its updated state, and so on. Although it would be more appropriate to use the term state-feedback, given that S_c has access to the states of S_a, we use feedback for brevity. To emphasize this feedback interpretation, the interconnection relation R^e is denoted by $\mathcal{F}$.

The next example illustrates the notion of feedback composition.

Example 6.2. Consider the system S_a displayed in Figure 6.1 and assume that we want to eliminate all the internal behaviors containing transitions of the form $x_{a1} \xrightarrow[a]{\mathtt{a}} x_{a1}$ or containing transitions of the form $x_{a0} \xrightarrow[a]{\mathtt{b}} x_{a0}$. This objective can be achieved by resorting to the controller S_c also represented in Figure 6.1. The required alternating simulation relation is given by:

$$\{(x_{c0}, x_{a0}), (x_{c1}, x_{a1}), (x_{c2}, x_{a2})\}.$$

The feedback composed system $S_c \times_{\mathcal{F}} S_a$ is also depicted in Figure 6.1 and it can be seen that it is equal, up to a relabeling of states and inputs, to S_c. Therefore, the controller enforces the desired requirements on S_a.

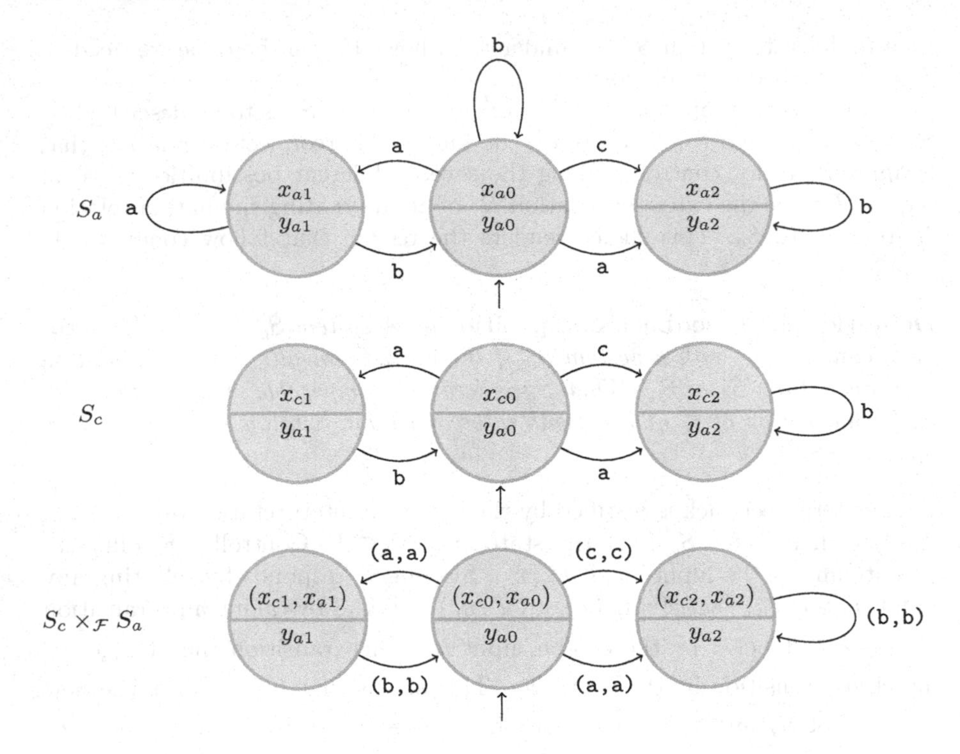

Fig. 6.1. From top to bottom we have: system S_a, controller S_c, and the feedback composed system $S_c \times_{\mathcal{F}} S_a$.

To understand the need to require the existence of an alternating simulation relation from S_c to S_a, let us attempt to use system S_d in Figure 6.2 as a controller. The relation $R = \{(x_{d0}, x_{a0}), (x_{d1}, x_{a1})\}$ is an obvious simulation relation from S_d to S_a but not an alternating simulation relation. Although the composition $S_d \times_{\mathcal{I}} S_a$ is well defined for the interconnection relation:

$$\mathcal{I} = \{(x_d, x_a, u_d, u_a) \in X_d \times X_a \times U_d \times U_a \mid (x_d, x_a) \in R\},$$

the transition $(x_{d0}, x_{a0}) \xrightarrow[da]{a,a} (x_{d1}, x_{a2})$ is not present in $S_d \times_{\mathcal{I}} S_a$ even though it is labeled by the same input as the transition $(x_{d0}, x_{a0}) \xrightarrow[da]{a,a} (x_{d1}, x_{a1})$ which is present in $S_d \times_{\mathcal{I}} S_a$. This means that an implementation of $S_d \times_{\mathcal{I}} S_a$ requires a synchronization procedure between S_d and S_a that is not purely based on inputs. $\lhd$

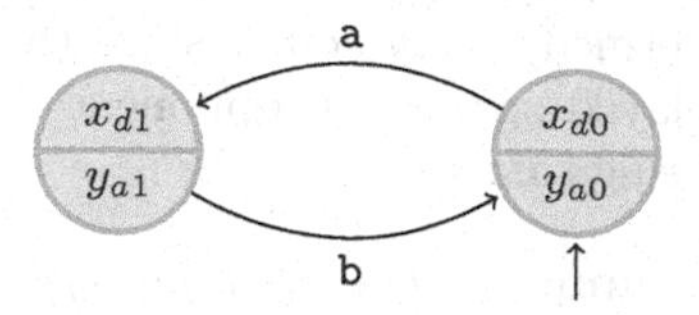

Fig. 6.2. Candidate controller for system S_a in Figure 6.1.

The following result, which is also valid for other forms of composition, explains how feedback composition can restrict the behavior of systems.

Proposition 6.3. *Let S_a and S_b be systems with $Y_a = Y_b$ and let $\mathcal{I}$ be an interconnection relation satisfying:*

$$(x_a, x_b) \in \pi_X(\mathcal{I}) \implies H_a(x_a) = H_b(x_b).$$

Then, the following holds:

- $S_a \times_{\mathcal{I}} S_b \preceq_{\mathcal{S}} S_b$;
- $S_b \times_{\mathcal{I}} S_a \preceq_{\mathcal{S}} S_a$.

Proof. The proof consists in routinely checking that the relations:

$$\{((x_a, x_b), x_b') \in X_{ab} \times X_b \mid x_b = x_b'\}$$
$$\{((x_b, x_a), x_a') \in X_{ba} \times X_a \mid x_a = x_a'\}$$

are simulation relations from $S_a \times_{\mathcal{I}} S_b$ to S_b and from $S_b \times_{\mathcal{I}} S_a$ to S_a, respectively. □

Feedback composition not only restricts the behavior of the system to be controlled but also its initial states. Recall that (x_c, x_a) is an initial state of $S_c \times_{\mathcal{F}} S_a$ if x_c is an initial state of S_c, x_a is an initial state of S_a, and $(x_c, x_a) \in R$. Therefore, it suffices that R does not relate x_a to an initial state of S_c to prevent x_a from being part of an initial state of $S_c \times_{\mathcal{F}} S_a$. The introduced notion of feedback composition thus assumes that the controller has the possibility of initializing S_a. As this assumption may not hold in many situations, we also show how to generalize the results in this chapter to the case where S_a cannot be initialized.

6.2 Safety games

We start by considering a very simple class of control problems whose objective is to design a controller S_c for a system S_a so that $S_c \times_{\mathcal{F}} S_a$ is nonblocking and $\mathrm{Reach}(S_c \times_{\mathcal{F}} S_a) \subseteq W$ for some set $W \subseteq Y_a$. If we regard W as a set of *safe* outputs, the objective of S_c is then to render W invariant for the behaviors in $\mathcal{B}^\omega(S_c \times_{\mathcal{F}} S_a)$, thus keeping the composed system safe. This class of control problems are termed *safety games* since the controller S_c arises as the solution of a game played against an opponent that tries to prevent the composed system from being safe.

Definition 6.4 (Safety game). *Let S_a be a system with $Y_a = X_a$ and $H_a = 1_{X_a}$, and let $W \subseteq X_a$ be a set of safe states. The* safety game *for system S_a and specification set W asks for the existence of a controller S_c such that:*

1. *S_c is feedback composable with S_a;*
2. *$S_c \times_{\mathcal{F}} S_a$ is nonblocking;*
3. *$\varnothing \neq \mathcal{B}^{\omega}(S_c \times_{\mathcal{F}} S_a) \subseteq W^{\omega}$.*

A safety game is said to be solvable when S_c exists.

The requirement $Y_a = X_a$ and $H_a = 1_{X_a}$ is made without loss of generality since the general case where $Y_a \neq X_a$ can be reduced to this one. We shall elaborate on this fact once we know how to solve safety games. Note that the third requirement in the preceding definition is equivalent to $\text{Reach}(S_c \times_{\mathcal{F}} S_a) \subseteq W$ since a behavior $y_0 y_1 y_2 \ldots$ in W^{ω} necessarily satisfies $y_i \in W$ for every $i \in \mathbb{N}_0$ and vice-versa.

Safety games can be solved by constructing a suitable operator:

$$F_W : 2^{X_a} \to 2^{X_a}$$

for any specification set $W \subseteq X_a$. A fixed-point of this operator provides a collection of states from which it is possible to control system S_a so as to remain in W. The operator F_W:

$$F_W(Z) = \{x_a \in Z \mid x_a \in W \text{ and } \exists u_a \in U_a(x_a) \quad \varnothing \neq \text{Post}_{u_a}(x_a) \subseteq Z\}$$

captures the essence of safety games in the sense that the set $F_W(Z)$ contains all the states $x_a \in Z \cap W$ for which all the u_a-successors of x_a are in Z. The next result shows that a maximal fixed-point of F_W exists and relates the solvability of safety games to fixed-points of F_W.

Proposition 6.5. *Let S_a be a system with $Y_a = X_a$ and $H_a = 1_{X_a}$, and let $W \subseteq X_a$ be a set of safe states. The operator $F_W : 2^X \to 2^X$ satisfies:*

1. *$Z \subseteq Z'$ implies $F_W(Z) \subseteq F_W(Z')$;*
2. *if the safety game for system S_a and specification set W is solvable, then the maximal fixed-point Z of F_W satisfies $Z \cap X_{a0} \neq \varnothing$.*

Proof. The first assertion follows directly from the definition of F_W.

To prove the second assertion, assume that a solution S_c to the safety game exists, let K be the set of all states reachable in $S_c \times_{\mathcal{F}} S_a$, and let $Z' = \text{Reach}(S_c \times_{\mathcal{F}} S_a)$. Note that $K \subseteq X_a \times X_b$ while $\pi_a(K) = Z' \subseteq X_a$. We claim that $X_{a0} \cap Z' \neq \varnothing$ and $Z' \subseteq F(Z')$. The first claim is proved by noting that the second and third requirement in the definition of safety game imply $K \cap X_{ca0} \neq \varnothing$ and thus:

$$Z' \cap X_{a0} = \pi_a(K) \cap \pi_a(X_{ca0}) \supseteq \pi_a(K \cap X_{ca0}) \neq \varnothing.$$

The second claim can be proved as follows. Let $x_a \in Z'$ and let $x_c \in X_c$ be such that $(x_c, x_a) \in K$. Since state (x_c, x_a) is reachable in $S_c \times_{\mathcal{F}} S_a$ and since S_c is a solution to the safety game, there must exist $(u_c, u_a) \in U_{ca}(x_c, x_a)$ such that $(x_c, x_a) \xrightarrow[ca]{u_c, u_a} (x'_c, x'_a)$ with $x'_a \in W$. Moreover, $(x'_c, x'_a) \in K$ which implies $x'_a \in Z'$. We now invoke the definition of feedback composition to conclude that every transition $x_a \xrightarrow[a]{u_a} x''_a$ in S_a labeled by the same input u_a gives rise to a transition $(x_c, x_a) \xrightarrow[ca]{u_c, u_a} (x''_c, x''_a)$ in $S_c \times_{\mathcal{F}} S_a$. Necessarily, $(x''_c, x''_a) \in K$ and thus $x''_a \in Z'$. Hence, we conclude the existence of an input $u_a \in U_a(x_a)$ for which $\varnothing \neq \mathrm{Post}_{u_a}(x_a) \subseteq Z'$. According to the definition of F_W, $x_a \in F_W(Z')$ and the second claim is proved. Finally, by definition of F_W we always have $F_W(Z') \subseteq Z'$ so that $F_W(Z') = Z'$ and Z' is a fixed-point of F_W. Since the second assertion in the proposition holds for any fixed-point Z' of F_W, it also holds for its maximal fixed-point whose existence is a consequence of the first assertion in the proposition. □

A controller solving a safety game with specification set W can always be constructed from the information contained in a fixed-point Z of F_W satisfying $Z \cap X_{a0} \neq \varnothing$. One possibility is the controller:

$$S_c = (X_c, X_{c0}, U_a, \xrightarrow[c]{}) \tag{6.1}$$

defined by:

- $X_c = Z$;
- $X_{c0} = Z \cap X_{a0}$;
- $x_c \xrightarrow[c]{u_a} x'_c$ if $\varnothing \neq \mathrm{Post}_{u_a}(x_c) \subseteq Z$,

and where $\mathrm{Post}_{u_a}(x_c)$ refers to the u_a-successors in S_a. It is a simple matter to check that the relation defined by all the pairs $(x_c, x_a) \in X_c \times X_a$ with $x_c = x_a$ is an alternating simulation relation from S_c to S_a. According to Proposition 6.3, $S_c \times_{\mathcal{F}} S_a \preceq_S S_c$ and we can interpret the result of composing S_c with S_a as the elimination of all the transitions labeled by inputs for which the corresponding successor sets are not contained in Z. Intuitively, S_c forces the behavior of $S_c \times_{\mathcal{F}} S_a$ to remain in $Z^\omega \subseteq W^\omega$.

A complete characterization of the solutions to safety games can now be obtained by noting that it follows from the results in the Appendix that the maximal fixed-point of F_W can be obtained by iterating F_W.

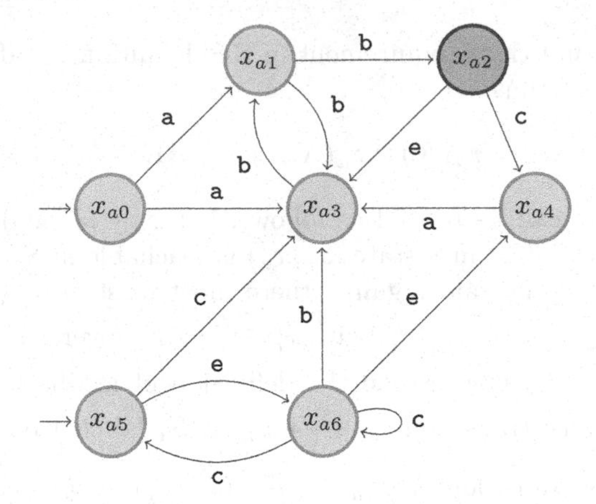

Fig. 6.3. System S_a for Example 6.7.

Theorem 6.6. *Let S_a be a system with $Y_a = X_a$ and $H_a = 1_{X_a}$, and let $W \subseteq X_a$ be a set of safe states. The safety game for system S_a and specification set W is solvable iff the maximal fixed-point Z of the operator F_W satisfies $Z \cap X_{a0} \neq \varnothing$. Moreover, Z can be obtained as:*

$$Z = \lim_{i \to \infty} F_W^i(X_a).$$

When $Z \cap X_{a0} \neq \varnothing$, a solution to the safety game is given by the controller (6.1).

Example 6.7. To illustrate Theorem 6.6 consider the finite-state system S_a in Figure 6.3 and let W be the set of all light-colored states. The maximal fixed-point of F_W can be obtained by iterating F_W and the result of this iteration is shown in Figure 6.4.

After 5 iterates of F_W a fixed-point is reached. The resulting set Z defines a controller that restricts the inputs to e at the state x_{a5} and to c at state x_{a6}. The reader can verify that this choice of inputs prevents the behavior of S_a to leave W^ω. The feedback composition of S_c with S_a, displayed in Figure 6.5, results in a finite-state system equal to S_c if we identify the states (x_{a5}, x_{a5}) and (x_{a6}, x_{a6}) with the states x_{a5} and x_{a6}, and if we identify the inputs (e, e) and (c, c) with the inputs e and c, respectively. ◁

The controller (6.1) is completely determined by a given fixed-point of F_W. When we use the maximal fixed-point of F_W, (6.1) becomes the best possible controller in the sense that any other controller solving the same safety problem would be more restrictive.

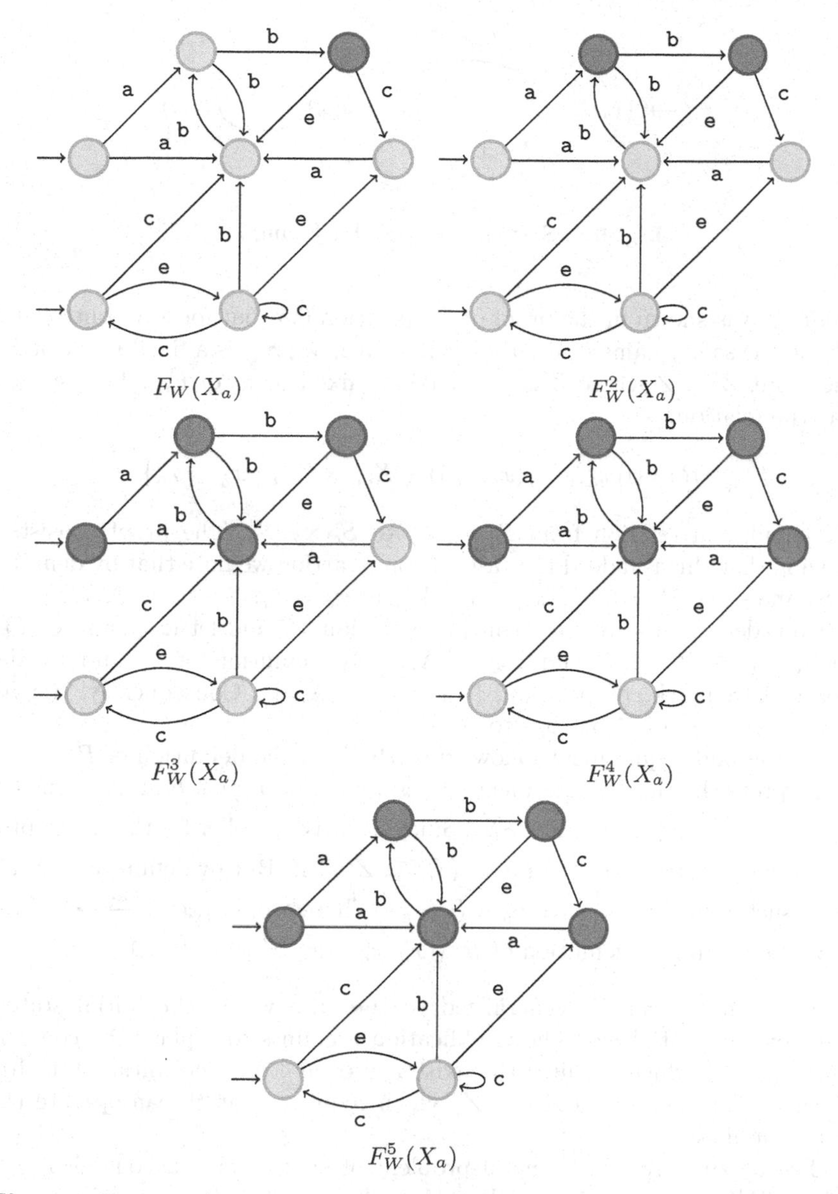

Fig. 6.4. Iterates of F_W. Dark-colored states correspond to the image of F_W.

Proposition 6.8. *Let S_a be a system with $Y_a = X_a$ and $H_a = 1_{X_a}$, and let $W \subseteq X_a$ be a set of safe states. For any controller S_d solving the safety game for system S_a and specification set W we have:*

$$S_d \times_{\mathcal{G}} S_a \preceq_{\mathcal{S}} S_c \times_{\mathcal{F}} S_a$$

where S_c is the controller (6.1) defined by the maximal fixed-point of F_W.

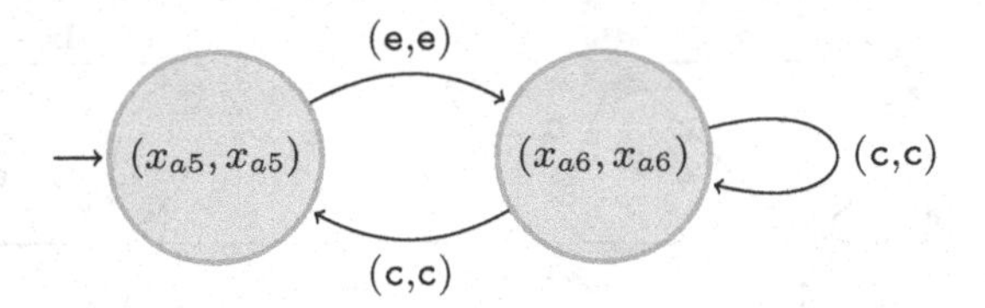

Fig. 6.5. System $S_c \times_{\mathcal{F}} S_a$ for Example 6.7.

Proof. It was shown in the proof of Proposition 6.5 that for any controller S_d solving the safety game, the set $Z' = \text{Reach}(S_d \times_{\mathcal{G}} S_a)$ is a fixed-point of F_W. Therefore, $Z' \subseteq Z$ where Z is the maximal fixed-point of F_W. This suggests that the relation:

$$R = \{((x_d, x'_a), (x_c, x_a)) \in X_{da} \times X_{ca} \mid x'_a = x_a\}$$

is a simulation relation from $S_d \times_{\mathcal{G}} S_a$ to $S_c \times_{\mathcal{F}} S_a$. The proof consists in showing that this is indeed the case. Before starting we note that by definition of S_c and since $H_a = 1_{X_a}$, $(x_c, x_a) \in X_{ca}$ iff $x_c = x_a$.

Consider the first requirement in Definition 4.7 and let $(x_{d0}, x_{a0}) \in X_{da0}$. Then, $x_{a0} \in Z' \subseteq Z$ and $x_{a0} \in X_{a0}$. By definition of S_c and by definition of feedback composition, $(x_{a0}, x_{a0}) \in X_{ca0}$. Consequently, the pair $((x_{d0}, x_{a0}), (x_{a0}, x_{a0}))$ belongs to R.

The second requirement follows directly from the definition of R.

To prove the third requirement let $((x_d, x_a), (x_a, x_a)) \in R$ and assume that $(x_d, x_a) \xrightarrow[da]{u_d, u_a} (x'_d, x'_a)$ in $S_d \times_{\mathcal{G}} S_a$. Since S_d is a controller for the safety problem we necessarily have $\varnothing \neq \text{Post}_{u_a}(x_a) \subseteq Z' \subseteq Z$. But by definition of S_c, for every such input u_a we have $u_a \in U_c(x_a)$. Therefore, $(x_a, x_a) \xrightarrow[ca]{u_a, u_a} (x'_a, x'_a)$ in $S_c \times_{\mathcal{F}} S_a$ and by definition of R, $((x'_d, x'_a), (x'_a, x'_a)) \in R$. □

Theorem 6.6 can be generalized to the case where the initial state of S_a cannot be initialized. The modification amounts to replace the condition $Z \cap X_{a0} \neq \varnothing$, which requires the existence of at least one initial state from which S_c can operate, to $X_{a0} \subseteq Z$, which requires that S_c can operate from every initial state in X_{a0}.

The apparently more general problem of synthesizing a controller S_c to enforce $\mathcal{B}^\omega(S_c \times_{\mathcal{F}} S_a) \subseteq W^\omega$ when $Y_a \neq X_a$ and $H_a \neq 1_{X_a}$ can be reduced to the problem studied in this section. It suffices to consider a new safe set $W' \subseteq X_a$ defined by $W' = H_a^{-1}(W)$ and to apply Theorem 6.6 to system $(X_a, X_{a0}, U_a, \xrightarrow[a]{}, X_a, 1_{X_a})$ and specification set W'.

6.3 Reachability games

While the objective of safety games is to keep the behaviors of the composed system within a safe set, reachability games ask for a certain set W of outputs to be reached. As in the previous section we consider only the case where $H_a = 1_{X_a}$ since the general case can be reduced to this one by suitably redefining W.

Definition 6.9 (Reachability game). *Let S_a be a system satisfying $Y_a = X_a$ and $H_a = 1_{X_a}$, and let $W \subseteq X_a$ be a set of states. The* reachability game *for system S_a and specification set W asks for the existence of a controller S_c such that:*

1. *S_c is feedback composable with S_a;*
2. *for every maximal behavior $\mathsf{y} \in \mathcal{B}(S_c \times_{\mathcal{F}} S_a) \cup \mathcal{B}^{\omega}(S_c \times_{\mathcal{F}} S_a)$ there exists $k \in \mathbb{N}_0$ such that $\mathsf{y}(k) = y_k \in W$.*

A reachability game is said to be solvable when S_c exists.

The second condition in the definition of reachability game requires that any infinite behavior $\mathsf{y} = y_0 y_1 \ldots$ of $S_c \times_{\mathcal{F}} S_a$ visits the set W in finite time. Moreover, it also requires that any finite behavior $y_0 y_1 \ldots y_l$ that cannot be extended to an infinite behavior, visits W before or when reaching a blocking state. In particular, no nonblocking condition is imposed since the objective is simply to reach W in finitely many steps. Once states in W are reached, no further requirements are imposed by the reachability game. More demanding requirements, such as reaching a set of states W and remaining within W thereafter, can be obtained by combining safety with reachability.

Similarly to safety games, reachability games can also be given a fixed-point characterization. For any $W \subseteq X_a$ we can define the operator:

$$G_W : 2^{X_a} \to 2^{X_a}$$

by:

$$G_W(Z) = \{x_a \in X_a \mid x_a \in W \text{ or } \exists u_a \in U_a(x_a) \quad \varnothing \neq \mathrm{Post}_{u_a}(x_a) \subseteq Z\}.$$

It is not difficult to see that for any $W \subseteq X_a$, the inclusion $Z \subseteq Z'$ implies $G_W(Z) \subseteq G_W(Z')$, thus guaranteeing the existence of a unique minimal fixed-point of G_W. Several different controllers solving the reachability game can be constructed from a fixed-point Z of G_W for which $Z \cap X_{a0} \neq \varnothing$.

Among the several possible solutions, we consider the controller:

$$S_c = (X_c, X_{c0}, U_a, \xrightarrow[c]{}) \tag{6.2}$$

defined as:

- $X_c = Z$;
- $X_{c0} = Z \cap X_{a0}$;
- $x_c \xrightarrow[c]{u_a} x'_c$ if there exists a $k \in \mathbb{N}$ such that $x_c \notin G_W^k(\varnothing)$ and $\varnothing \neq \mathrm{Post}_{u_a}(x_c) \subseteq G_W^k(\varnothing)$,

and where $\mathrm{Post}_{u_a}(x_c)$ refers to the u_a-successors in S_a. Moreover, one can easily verify that the relation defined by all the pairs $(x_c, x_a) \in X_c \times X_a$ with $x_c = x_a$ is an alternating simulation relation from S_c to S_a. Similarly to safety games, the solution of reachability games can be fully characterized in terms of the fixed-points of G_W.

Theorem 6.10. *Let S_a be a system with $Y_a = X_a$ and $H_a = 1_{X_a}$, and let $W \subseteq X_a$ be a set of states. The reachability game for S_a and specification set W is solvable iff the minimal fixed-point Z of the operator G_W satisfies $Z \cap X_{a0} \neq \varnothing$. Moreover, Z can be obtained as:*

$$Z = \lim_{i \to \infty} G_W^i(\varnothing).$$

When $Z \cap X_{a0} \neq \varnothing$, a solution to the reachability game is given by the controller (6.2).

Example 6.11. Consider again the finite-state system in Figure 6.3 and assume that W consists of the single state x_{a4}. The computation of the fixed-point of G_W by iteration is presented in Figure 6.6. The resulting controller (6.2) is displayed in Figure 6.7. Note that state x_{a2} is not helpful for this particular problem since it is not reachable. However, it may be useful when this set of states, corresponding to the minimal fixed-point of G_W, is used as the starting point for further design problems. $\lhd$

For reachability games there is no optimal controller in the sense of Proposition 6.8. This observation is illustrated in Example 6.12.

Example 6.12. Consider the system S_a in Figure 6.8 where the set W consists of the state x_2. Let S_c be any finite-state controller solving the reachability problem for system S_a and let k be the maximum number of times[2] that an internal behavior of $S_c \times_{\mathcal{F}} S_a$ visits the state x_1 before reaching the state x_2. We can always construct a controller S_d that allows the internal behaviors of $S_d \times_{\mathcal{G}} S_a$ to visit x_1 any number of times smaller than or equal to $k+1$ before reaching x_2. Clearly, S_d is less restrictive than S_c which shows that a minimally restrictive controller does not exist. $\lhd$

[2] Such number exists since both S_c and S_a are finite-state systems.

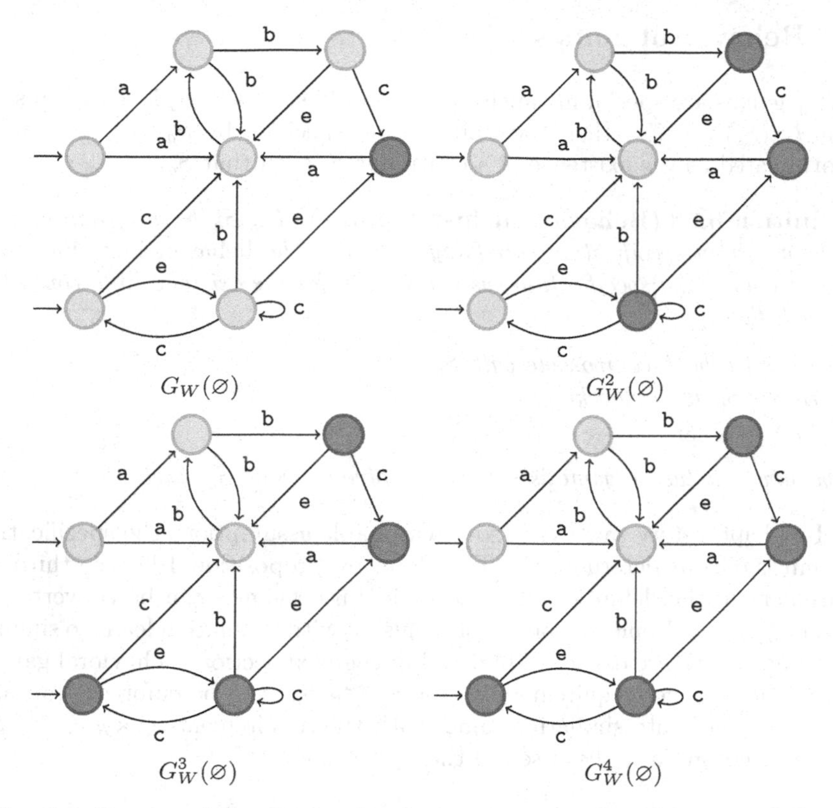

Fig. 6.6. Iterates of G_W. Dark-colored states correspond to the image of G_W.

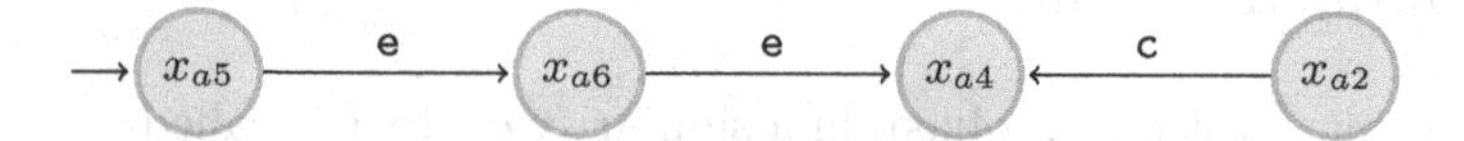

Fig. 6.7. Controller S_c for Example 6.11.

Analogously to safety games, Theorem 6.10 can be generalized to the case where the initial states of S_a cannot be initialized by the controller. This generalization consists in replacing $Z \cap X_{a0} \neq \varnothing$ with the stronger condition $X_{a0} \subseteq Z$ guaranteeing that no initial state of S_a is eliminated in the composition with the controller.

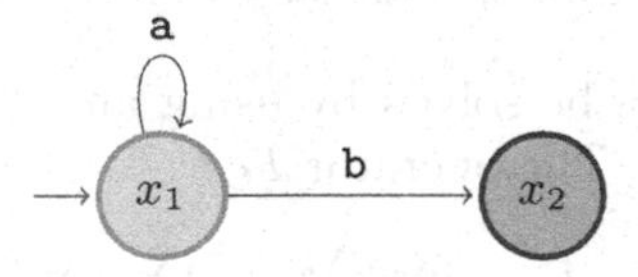

Fig. 6.8. System S_a for Example 6.12. The set W consists of the state x_2.

6.4 Behavioral games

Safety games are special instances of behavioral games. If S_b is a system such that $\mathcal{B}^\omega(S_b) = W^\omega$, then the safety game specified by S_a and W can be reformulated as the existence of a controller S_c such that $S_c \times_{\mathcal{F}} S_a \preceq_{\mathcal{B}} S_b$.

Definition 6.13 (Behavior inclusion game). *Let S_a be a system and let S_b be a system specification satisfying $Y_b = Y_a$. The* behavior inclusion game *for system S_a and specification system S_b asks for the existence of a controller S_c such that:*

1. *S_c is feedback composable with S_a;*
2. *$S_c \times_{\mathcal{F}} S_a$ is nonblocking;*
3. *$S_c \times_{\mathcal{F}} S_a \preceq_{\mathcal{B}} S_b$.*

A behavior inclusion game is said to be solvable when S_c exists.

In Chapter 4 we saw that under reasonable assumptions the specification system is output deterministic. Therefore, by Proposition 4.11, the third requirement in the definition of behavior inclusion games can be converted to $S_c \times_{\mathcal{F}} S_a \preceq_{\mathcal{S}} S_b$. Replacing behavior inclusion with simulation leads to similarity games which are discussed in detail in the next section. Behavioral games, where the stronger requirement $S_c \times_{\mathcal{F}} S_a \cong_{\mathcal{B}} S_b$ is to be enforced, can also be transformed into similarity games with the requirement $S_c \times_{\mathcal{F}} S_a \cong_{\mathcal{S}} S_b$. These games are also discussed in the next section.

6.5 Similarity games

The controller synthesis problem in a similarity context is called a *simulation game.*

Definition 6.14 (Simulation game). *Let S_a be a system and let S_b be a system specification satisfying $Y_b = Y_a$. The* simulation game *for system S_a and specification system S_b asks for the existence of a controller S_c such that:*

1. *S_c is feedback composable with S_a;*
2. *$S_c \times_{\mathcal{F}} S_a$ is nonblocking;*
3. *$S_c \times_{\mathcal{F}} S_a \preceq_{\mathcal{S}} S_b$.*

A simulation game is said to be solvable when S_c exists.

Simulation games can be solved by using an extension of the operator F introduced in Chapter 5. The operator F_C:

$$F_C : 2^{X_a \times X_b} \to 2^{X_a \times X_b}$$

in which the subscript C refers to control, is defined by $(x_a, x_b) \in F_C(W)$, for some $W \subseteq X_a \times X_b$, if the following three conditions are satisfied:

1. $H_a(x_a) = H_b(x_b)$;
2. $(x_a, x_b) \in W$;
3. there exists $u_a \in U_a(x_a)$ such that for every $x'_a \in \mathrm{Post}_{u_a}(x_a)$ there exists a transition $x_b \xrightarrow[b]{u_b} x'_b$ in S_b with $(x'_a, x'_b) \in W$.

As before, $Z \subseteq Z'$ implies $F_C(Z) \subseteq F_C(Z')$ so that F_C has a unique maximal fixed-point which can be used to construct a solution to the simulation game whenever $Z \cap (X_{a0} \times X_{b0}) \neq \varnothing$. In this case we can define the controller:

$$S_c = (X_c, X_{c0}, U_a, \xrightarrow[c]{}, Y_a, H_c) \tag{6.3}$$

by:

- $X_c = Z$;
- $X_{c0} = Z \cap (X_{a0} \times X_{b0})$;
- $(x_a, x_b) \xrightarrow[c]{u_a} (x'_a, x'_b)$ in S_c if the following three conditions hold:
  1. $(x'_a, x'_b) \in Z$;
  2. $x_b \xrightarrow[b]{u_b} x'_b$ in S_b for some $u_b \in U_b(x_b)$;
  3. $x_a \xrightarrow[a]{u_a} x'_a$ in S_a for some $u_a \in U_a(x_a)$ such that for all $x''_a \in \mathrm{Post}_{u_a}(x_a)$ there exists a transition $x_b \xrightarrow[b]{u'_b} x''_b$ in S_b with $(x''_a, x''_b) \in Z$;
- $H_c(x_a, x_b) = H_a(x_a)$.

The reader should verify that the definition of F_C ensures that the relation:

$$R = \{((x_a, x_b), x'_a) \in Z \times X_a \mid x_a = x'_a\}$$

is an alternating simulation relation from S_c to S_a.

The previous discussion can be summarized in the following result characterizing the solution to behavior inclusion games.

Theorem 6.15. *Let S_a be a system and let S_b be a system specification with $Y_b = Y_a$. The simulation game for system S_a and specification system S_b is solvable iff the maximal fixed-point Z of the operator F_C satisfies $Z \cap (X_{a0} \times X_{b0}) \neq \varnothing$. Moreover, Z can be obtained as:*

$$Z = \lim_{i \to \infty} F_C^i(X_a \times X_b).$$

When $Z \cap (X_{a0} \times X_{b0}) \neq \varnothing$, a solution to the simulation game is given by the controller (6.3).

Proof. It was already shown, by defining explicitly the controller S_c in (6.3), that existence of a fixed-point Z of F_C satisfying $Z \cap (X_{a0} \times X_{b0}) \neq \varnothing$ leads to a solution of the simulation game.

The converse implication can be proved by noting that from any controller S_c solving the simulation game and from the corresponding simulation relation R from $S_c \times_{\mathcal{F}} S_a$ to S_b we can construct a relation $R' \subseteq X_a \times X_b$ defined by $(x_a, x_b) \in R'$ if there exists $x_c \in X_c$ such that $((x_c, x_a), x_b) \in R$. It is now simple to verify that R' is a fixed-point of F_C. The crucial inclusion is $R' \subseteq F_C(R')$ and the key step is to show that any $(x_a, x_b) \in R'$ satisfies the third requirement in the definition of F_C. We focus on this step. Let $(x_a, x_b) \in R'$ and recall that by definition of R' there exists $x_c \in X_c$ such that $((x_c, x_a), x_b) \in R$. Since $S_c \times_{\mathcal{F}} S_a$ is nonblocking, there exists an input $(u_c, u_a) \in U_{ca}(x_c, x_a)$. Moreover, it follows from the definition of feedback composition, that for every $x'_a \in \mathrm{Post}_{u_a}(x_a)$ there exists a transition $(x_c, x_a) \xrightarrow[ca]{u_c, u_a} (x'_c, x'_a)$ in $S_c \times_{\mathcal{F}} S_a$. But as R is a simulation relation from $S_c \times_{\mathcal{F}} S_a$ to S_b, for every transition $(x_c, x_a) \xrightarrow[ca]{u_c, u_a} (x'_c, x'_a)$ in $S_c \times_{\mathcal{F}} S_a$ there exists a transition $x_b \xrightarrow{u'_b} x'_b$ in S_b satisfying $((x'_c, x'_a), x'_b) \in R$. We thus conclude the existence of $u_a \in U_a(x_a)$ such that for every $x'_a \in \mathrm{Post}_{u_a}(x_a)$ there exists a transition $x_b \xrightarrow{u'_b} x'_b$ in S_b satisfying $(x'_a, x'_b) \in R'$ which is precisely the third requirement in the definition of F_C. □

Example 6.16. To illustrate the construction of S_c we revisit the models for the bus fare machine used in Example 4.3 and displayed in Figure 4.1 and Figure 4.2. Although $S_a \cong_{\mathcal{B}} S_b$, system S_a is not simulated by system S_b. We thus seek a controller solving the simulation game for system S_a and specification system S_b. The iteration of F_C starts with the set $X_a \times X_b$ and terminates with the fixed-point Z after two iterations.

$$
\begin{aligned}
F_C^0(X_a \times X_b) &= \{(x_{a0}, x_{b0}), (x_{a0}, x_{b1}), (x_{a0}, x_{b2}), (x_{a0}, x_{b3}), (x_{a1}, x_{b0}),\\
&\quad (x_{a1}, x_{b1}), (x_{a1}, x_{b2}), (x_{a1}, x_{b3}), (x_{a2}, x_{b0}), (x_{a2}, x_{b1}),\\
&\quad (x_{a2}, x_{b2}), (x_{a2}, x_{b3})\},\\
F_C^1(X_a \times X_b) &= \{(x_{a0}, x_{b0}), (x_{a1}, x_{b1}), (x_{a1}, x_{b3}), (x_{a2}, x_{b2})\},\\
F_C^2(X_a \times X_b) &= \{(x_{a0}, x_{b0}), (x_{a1}, x_{b1}), (x_{a1}, x_{b3}), (x_{a2}, x_{b2})\}.
\end{aligned}
$$

The corresponding controller S_c is displayed in Figure 6.9 and $S_c \times_{\mathcal{F}} S_a$ is shown in Figure 6.10. The simulation relation from $S_c \times_{\mathcal{F}} S_a$ to S_b is given by the fixed-point Z of F_C. Note that the state (x_{a1}, x_{b3}), albeit not reachable, can be useful when $S_c \times_{\mathcal{F}} S_a$ is the starting point for further designs. ◁

The operator F_C can be seen as a control generalization of the operator F defined in Chapter 5. If there exists a simulation relation from system S_a to system S_b, then the maximal fixed-point F_C coincides with the maximal fixed-point of F. However, if no simulation relation from S_a to S_b exists, then

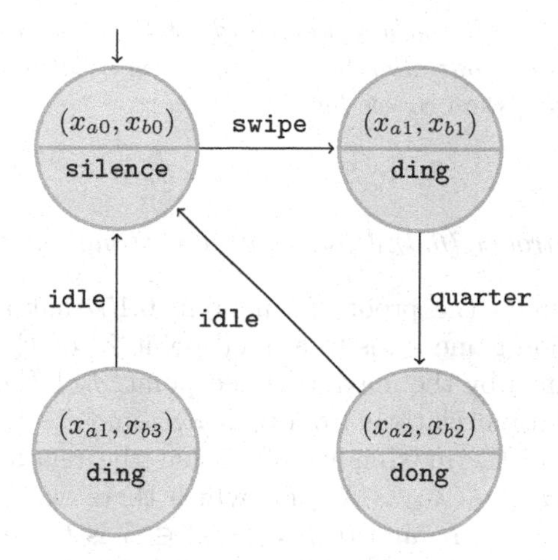

Fig. 6.9. Controller S_c for Example 6.16.

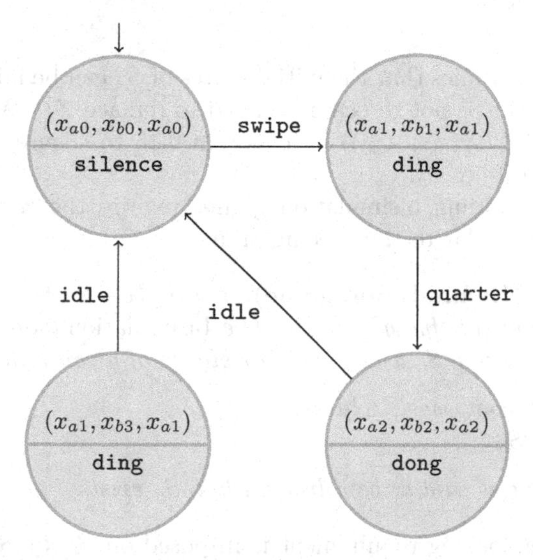

Fig. 6.10. Composed system $S_c \times_{\mathcal{F}} S_a$ for Example 6.16.

$X_{a0} \not\subseteq \pi_a(Z \cap (X_{a0} \times X_{b0}))$ for the maximal fixed-point Z of F. In contrast, the maximal fixed-point of F_C provides a simulation relation from a restricted version of S_a to S_b whenever a solution for the simulation game exists. Such restricted version can then be constructed as $S_c \times_{\mathcal{F}} S_a$ for a controller S_c.

The iteration of the operator F_C provides an algorithm which is guaranteed to terminate in time polynomial in $|X_a||X_b|$ for finite-state systems. Moreover, the controller S_c constructed from the maximal fixed-point of F_C in (6.3) is optimal in the sense that it minimally restricts S_a in order to enforce the specification.

Proposition 6.17. *Let S_a be a system and let S_b be a system specification with $Y_b = Y_a$. For any controller S_d solving the simulation game for system S_a and specification system S_b we have:*

$$S_d \times_{\mathcal{G}} S_a \preceq_{\mathcal{S}} S_c \times_{\mathcal{F}} S_a$$

where S_c is the controller (6.3) defined by the maximal fixed-point of F_C.

Proof. It was shown in the proof of Theorem 6.15 that any controller S_d solving the simulation game leads to a fixed-point Z' of F_C. Moreover, S_c is completely determined by the maximal fixed-point Z of F_C. If we denote by R and R' the simulation relations from $S_c \times_{\mathcal{F}} S_a$ and $S_d \times_{\mathcal{G}} S_a$, respectively, to S_b, it follows from the maximality of Z that the relation defined by the pairs $((x_d, x'_a), (x_c, x_a)) \in X_{da} \times X_{ca}$ for which there exists a state $x_b \in X_b$ such that $((x_d, x'_a), x_b) \in R'$ and $((x_c, x_a), x_b) \in R$ is the desired simulation relation from $S_d \times_{\mathcal{G}} S_a$ to $S_c \times_{\mathcal{F}} S_a$. The rest of the proof consists in routinely checking that all the requirements in Definition 4.7 are satisfied and is left to the reader. □

Theorem 6.15 assumes that the initial states of S_a can be initialized by the controller. When this is not the case we need to replace $Z \cap (X_{a0} \times X_{b0}) \neq \varnothing$ with $X_{a0} = \pi_a(Z \cap (X_{a0} \times X_{b0}))$ in Theorem 6.15 to ensure S_c can operate from any initial state of S_a.

The more demanding bisimulation games require the composed system $S_c \times_{\mathcal{F}} S_a$ to be bisimilar to the specification.

Definition 6.18 (Bisimulation game). *Let S_a be a system and let S_b be a system specification satisfying $Y_b = Y_a$. The* bisimulation game *for system S_a and specification system S_b asks for the existence of a controller S_c such that:*

1. *S_c is feedback composable with S_a;*
2. *$S_c \times_{\mathcal{F}} S_a \cong_{\mathcal{S}} S_b$.*

A simulation game is said to be solvable when S_c exists.

Note that no nonblocking requirement is imposed on $S_c \times_{\mathcal{F}} S_a$ since a state in $S_c \times_{\mathcal{F}} S_a$ is blocking iff it is bisimulated by a blocking state in S_b. Hence, the existence or absence of blocking states is completely determined by the specification S_b.

Before molding bisimulation games into a fixed-point computation we make two observations. First, if there exists a controller S_c rendering $S_c \times_{\mathcal{F}} S_a$ bisimilar to S_b, it follows that $S_b \preceq_{\mathcal{S}} S_a$ since $S_b \cong_{\mathcal{S}} S_c \times_{\mathcal{F}} S_a \preceq_{\mathcal{S}} S_a$ in virtue of Proposition 6.3. The second observation notes that for any input $u_a \in U_a(x_a)$ enabled by the controller S_c and for any $x'_a \in \mathrm{Post}_{u_a}(x_a)$ there must exist a matching transition $x_b \xrightarrow[b]{u_b} x'_b$ in S_b.

The preceding observations motivate the definition of the operator:

$$G_C : 2^{X_a \times X_b} \to 2^{X_a \times X_b}$$

given by $(x_a, x_b) \in G_C(W)$, for some $W \subseteq X_a \times X_b$, if the following three conditions are satisfied:

1. $H_a(x_a) = H_b(x_b)$;
2. $(x_a, x_b) \in W$;
3. for every transition $x_b \xrightarrow[b]{u_b} x'_b$ in S_b there exists an input $u_a \in U_a(x_a)$ satisfying:
   a) there exists $x'_a \in \text{Post}_{u_a}(x_a)$ with $(x'_a, x'_b) \in W$;
   b) for every $x''_a \in \text{Post}_{u_a}(x_a)$ there exists a transition $x_b \xrightarrow[b]{u'_b} x''_b$ in S_b with $(x''_a, x''_b) \in W$.

A controller based on a fixed-point Z of G_C can be constructed whenever $\pi_b(Z \cap (X_{a0} \times X_{b0})) = X_{b0}$. Under this assumption, one possible controller is:

$$S_c = (X_c, X_{c0}, U_a, \xrightarrow[c]{}, Y_a, H_c) \tag{6.4}$$

defined by:

- $X_c = Z$;
- $X_{c0} = Z \cap (X_{a0} \times X_{b0})$;
- for every transition $x_b \xrightarrow[b]{u_b} x'_b$ in S_b, $(x_a, x_b) \xrightarrow[c]{u_a} (x'_a, x'_b)$ in S_c if the following two conditions hold:
  1. $(x'_a, x'_b) \in Z$;
  2. $x_a \xrightarrow[a]{u_a} x'_a$ in S_a for some $u_a \in U_a(x_a)$ such that for all $x''_a \in \text{Post}_{u_a}(x_a)$ there exists a transition $x_b \xrightarrow[b]{u'_b} x''_b$ in S_b with $(x''_a, x''_b) \in Z$;
- $H_c(x_a, x_b) = H_a(x_a)$.

Controller S_c is defined so as to make the relation:

$$\{((x_a, x_b), x'_a) \in Z \times X_a \mid x_a = x'_a\}$$

an alternating simulation relation from S_c to S_a. Arguing as we did for simulation games we arrive at the following result characterizing the solution of bisimulation games.

Theorem 6.19. *Let S_a be a system and let S_b be a system specification with $Y_b = Y_a$. The bisimulation game for system S_a and specification system S_b is solvable iff the maximal fixed-point Z of the operator G_C satisfies $\pi_b(Z \cap (X_{a0} \times X_{b0})) = X_{b0}$. Moreover, Z can be obtained as:*

$$Z = \lim_{i \to \infty} G_C^i(X_a \times X_b).$$

When $\pi_b(Z \cap (X_{a0} \times X_{b0})) = X_{b0}$, a solution to the simulation game is given by the controller (6.4).

For finite-state systems, a fixed-point of G_C is reached after finitely many iterations and the bisimulation game is solvable in time polynomial in $|X_a||X_b|$. Although the operator G_C can also be used for infinite-state systems, a fixed-point may not be reached in finitely many steps unless one is working with an infinite-state system satisfying additional assumptions such as the ones described in Part III.

In situations where it is not possible to initialize S_a we can still apply Theorem 6.19 by strengthening it with the requirement $\pi_a(Z \cap (X_{a0} \times X_{b0})) = X_{a0}$.

6.6 Notes

Problems of control in the behavioral context have been studied in the discrete-event systems community since the pioneering work of Ramadge and Wonham [RW87, RW89]. The main results of this line of work can now be found in several books [KG95, CL99]. Similar problems were independently solved in the context of reactive software synthesis [PR89a, PR89b]. Except for [QL91], the corresponding simulation and bisimulation games have been addressed much more recently and using very different mathematical formalizations [MT02, AVW03, Tab04, ZKJ06, Tab08b]. The use of alternating simulation relations to formalize feedback composition and the systematic exposition based on fixed-points appears to be new.

The adroit reader certainly noticed the reachability problem to be different from all the other control problems considered in this chapter: its solution is given by a minimal and not a maximal fixed-point, and no least restrictive controller exists. Reachability is an instance of a liveness property as opposed to safety. See, *e.g.*, [AS87] for definitions of safety and liveness. This distinction between safety and liveness properties also occurs in verification problems and makes verification a much more interesting topic than what can be judged by the superficial treatment in Chapter 5.

Worth mentioning is also the similarity between the definition of the operator G_C, used to solve bisimulation games, and the definition of alternating simulation relation. This is no coincidence since the solutions of bisimulation games can be completely characterized in terms of certain alternating simulation relations, see [Tab04, Tab08b].

Part III

Infinite Systems: Exact symbolic models

7

Exact symbolic models for verification

The evolution of physical quantities such as position, temperature, humidity, etc, is usually described by differential equations with solutions evolving on $\mathbb{R}^n$ or appropriate subsets. The infinite cardinality of $\mathbb{R}^n$ prevents a direct application of the verification methods described in Chapter 5. However, verification algorithms are still applicable whenever suitable finite-state abstractions of these infinite-state systems can be constructed. In recent years, several methods have been proposed for the construction of these abstractions based on a very interesting blend of different mathematical techniques. We present several of these methods starting with timed automata to illustrate the general principles of the abstraction process. Most of the abstraction techniques described in this chapter require linear differential equations. For this reason, we discuss as a special topic how to transform a class of nonlinear differential equations into linear differential equations in larger state spaces.

Notation

The identity map on a set Z is denoted by $1_Z : Z \to Z$. The graph of a function $f : Z \to W$ is denoted by $\Gamma(f)$ and defined as the set $\Gamma(f) = \{(z, w) \in Z \times W \mid w = f(z)\}$. The pre-image of $K \subseteq W$ under f is the set $f^{-1}(K) = \{z \in Z \mid f(z) \in K\}$. The function sign : $\mathbb{R} \to \{-1, 0, 1\}$ is defined by $\text{sign}(x) = -1$ for $x < 0$, $\text{sign}(x) = 0$ for $x = 0$, and $\text{sign}(x) = 1$ for $x > 0$.

Given an equivalence relation Q on a set Z, we denote by $[z]$ the equivalence class of $z \in Z$, by Z/Q the set of all equivalence classes, and by $\pi_Q : Z \to Z/Q$ the natural projection map taking a point $z \in Z$ to its equivalence class $\pi(z) = [z] \in Z/Q$. We say that an equivalence relation is finite when it has finitely many equivalence classes. An equivalence relation Q refines an equivalence relation R when $(z, z') \in Q$ implies $(z, z') \in R$. Given a collection of sets $\mathcal{Z} = \{Z_i\}_{i \in I}$, $Z_i \subseteq Z$, we say that an equivalence relation Q on Z respects $\mathcal{Z}$ if $(z, z') \in Q$ implies $z \in Z_i$ iff $z' \in Z_i$ for every $i \in I$.

P. Tabuada, *Verification and Control of Hybrid Systems: A Symbolic Approach*,
DOI: 10.1007/978-1-4419-0224-5_7, © Springer Science + Business Media, LLC 2009

A finite partition $\mathcal{P}$ of a set Z is a finite collection of sets $\mathcal{P} = \{P_i\}_{i\in I}$, $P_i \subseteq Z$, satisfying $\cup_{i\in I} P_i = Z$ and $i \neq j \implies P_i \cap P_j = \varnothing$ for any $i, j \in I$.

For any set $Z \subseteq \mathbb{R}^n$, $\overline{Z}$ denotes the topological closure of Z and $\operatorname{int} Z$ denotes its interior. The Minkowski sum of two sets $Z, W \subseteq \mathbb{R}^n$ is the set $Z + W = \{x \in \mathbb{R}^n \mid x = z + w \text{ for some } z \in Z \text{ and } w \in W\}$. The infinity norm of a vector $x \in \mathbb{R}^n$ is denoted by $\|x\|_\infty = \max_{i=1,\dots,n} |x_i|$ and the corresponding induced matrix norm by $\|A\|_\infty = \max_{i=1,\dots,m} \sum_{j=1}^{n} |a_{ij}|$ for $A \in \mathbb{R}^{n\times m}$ where a_{ij} is the entry of matrix A on the ith row and the jth column. The natural projections taking a vector $x \in \mathbb{R}^n$ to its ith component x_i are denoted by $\pi_i : \mathbb{R}^n \to \mathbb{R}$, $i = 1, 2, \dots, n$.

7.1 Dynamical and hybrid dynamical systems as systems

With the objective of constructing finite-state abstractions for verification, we show in this chapter how to model dynamical systems and hybrid dynamical systems as systems.

7.1.1 Dynamical systems

The reader less familiar with differential equations will benefit from rereading the rigid body example in Chapter 1 before progressing through this section.

We consider differential equations:

$$\frac{d}{dt}\xi = f(\xi) \tag{7.1}$$

in which $f : \mathbb{R}^n \to \mathbb{R}^n$ is an infinitely differentiable[1] function. We also call such functions smooth. The simpler notation $\dot{\xi} = f(\xi)$ is also used to denote (7.1). By a *solution* or *trajectory* of (7.1), with initial condition $x \in \mathbb{R}^n$, we mean a smooth curve:

$$\xi :]a, b[\to \mathbb{R}^n$$

satisfying:

1. $a < 0 < b$;
2. $\xi(0) = x$;
3. $\frac{d}{dt}\xi(t) = f(\xi(t))$ for all $t \in]a, b[$.

When we want to emphasize the initial condition $x \in \mathbb{R}^n$, we denote a trajectory by ξ_x. For the preceding class of differential equations, classical results guarantee that for any initial condition $x \in \mathbb{R}^n$ there exists a *unique* solution $\xi :]a_x, b_x[\to \mathbb{R}^n$ of (7.1) where the constants $a_x, b_x \in \mathbb{R}$ depend on x. When $-a_x = b_x = +\infty$ for any $x \in \mathbb{R}^n$, the differential equation is said to be

[1] Most results in this chapter hold under weaker regularity assumptions. The differentiability assumption is only made for simplicity of presentation.

complete. For complete differential equations there exists a unique solution defined for all $t \in \mathbb{R}$ and for every initial condition. We can thus speak of a family of solutions indexed by $x \in \mathbb{R}^n$. This family is denoted by:

$$\theta : \mathbb{R}^n \times \mathbb{R} \to \mathbb{R}^n$$

where $\theta(x,t) = \xi_x(t)$ is the solution of (7.1) with initial condition x. It is known from classical results on differential equations that θ satisfies:

$$\theta(x, 0) = x \tag{7.2}$$

$$\theta(x, t_1 + t_2) = \theta(\theta(x, t_1), t_2) \tag{7.3}$$

for very $x \in \mathbb{R}^n$ and $t_1, t_2 \in \mathbb{R}$. Moreover, if we are given a smooth map $\theta : \mathbb{R}^n \times \mathbb{R} \to \mathbb{R}^n$ satisfying (7.2) and (7.3), there is a unique differential equation that has $\theta(x,t)$ as its solution with initial condition x. For convenience, we will use both f and θ to define the corresponding differential equation. The pair $(\mathbb{R}^n, f)$ or $(\mathbb{R}^n, \theta)$ defines a *dynamical system.*

Definition 7.1 (Dynamical system). *A* dynamical system Σ *is a pair* $(\mathbb{R}^n, f)$ *consisting of :*

- *the state space* $\mathbb{R}^n$*;*
- *a smooth map* $f : \mathbb{R}^n \to \mathbb{R}^n$ *defining a differential equation as in (7.1).*

A solution *or* trajectory *of a dynamical system is a solution or trajectory of the differential equation defined by* f*. A dynamical system is said to be* complete *when the differential equation defined by* f *is complete.*

We will also consider dynamical systems $(\mathcal{X}, f)$ with f defined in a strict subset $\mathcal{X}$ of $\mathbb{R}^n$. We note that in such cases a solution is a curve $\xi :]a, b[\to \mathcal{X}$ with image in $\mathcal{X}$.

When $\Sigma = (\mathbb{R}^n, f)$ describes the evolution of some physical quantity being measured or regulated by a control system, we are many times confronted with the need to prove that θ or ξ satisfies some desired requirements or specifications. Moreover, the requirements to be proven seldom require that we distinguish between arbitrary elements of $\mathbb{R}^n$. In a chemical plant, for example, it is usually necessary to guarantee that pressures and concentrations are controlled to lie within a set of admissible values rather than to a specific value. We can thus construct a finite set of outputs:

$$Y = \{\texttt{Ok}, \texttt{TempViolation}, \texttt{ConcViolation}, \texttt{TempAndConcViolation}\}$$

and an output map $H : \mathbb{R}^n \to Y$ sending: the set of states in which temperature and concentration are within the desired range to the symbol `Ok`; the set of states in which temperature is outside the desired range but concentration is within the desired range to the symbol `TempViolation`; the set of states in which concentration is outside the desired range but temperature is within the desired range to the symbol `ConcViolation`; and the set of states in which both temperature and concentration are outside the desired range to the symbol `TempAndConcViolation`. The behavior of Σ observed through the output map H is a subset of Y^ω, and since Y is a finite set, it is natural to ask if there exists a finite-state system generating the same output behavior. The answer will depend not only on Σ but also on the choice of output set Y and output map H. Hence, we revisit the different ways in which we can specify Y and H and how such choices lead to a system describing Σ.

There are different equivalent ways in which we can define the output of a dynamical system Σ. As in the previous chemical plant example, we can specify an output map $H : \mathbb{R}^n \to Y$ to a finite set Y describing the desired outputs. Such map defines an equivalence relation Q on $\mathbb{R}^n$ by $(x, x') \in Q$ if $H(x) = H(x')$. Equivalently, we can directly specify an equivalence relation Q on $\mathbb{R}^n$ having a finite number of equivalence classes. The relation Q defines the canonical projection map $\pi_Q : \mathbb{R}^n \to \mathbb{R}^n/Q$ that can be used as output map. A third equivalent way consists in defining a finite partition $\mathcal{P}$ of $\mathbb{R}^n$. Given one such partition, we can regard each set $P \in \mathcal{P}$ as an equivalence class and obtain an equivalence relation. Conversely, an equivalence relation on $\mathbb{R}^n$ defines a partition of $\mathbb{R}^n$ where each set $P \in \mathcal{P}$ is an equivalence class of Q.

Since all the preceding ways of defining the output set and map are equivalent, we take a dynamical system to be verified and a finite equivalence relation on its state set as the starting point for the methods described in this chapter.

Definition 7.2. *Consider a dynamical system $\Sigma = (\mathbb{R}^n, f)$, let Q be a finite equivalence relation on $\mathbb{R}^n$, and let $L \subseteq \mathbb{R}^n$ be a set of initial states. The system associated with Σ, Q, and L, denoted by $S_{QL}(\Sigma)$, consists of:*

- $X = \mathbb{R}^n$;
- $X_0 = L$;
- $U = \mathbb{R}^+_0$;
- $x \xrightarrow{\tau} x'$ *if any of the following two conditions is satisfied:*
  1. $\pi_Q(x) \neq \pi_Q(x')$, $\xi_x : [0, \tau] \to \mathbb{R}^n$ *is a solution*[2] *of Σ satisfying* $\xi_x(\tau) = x'$, *and there exists* $\varepsilon \in [0, \tau]$ *satisfying one of the following:*
     a) $\pi_Q(\xi_x(t)) = \pi_Q(x)$ *for* $t \in [0, \varepsilon[$ *and* $\pi_Q(\xi_x(t)) = \pi_Q(x')$ *for* $t \in [\varepsilon, \tau]$;
     b) $\pi_Q(\xi_x(t)) = \pi_Q(x)$ *for* $t \in [0, \varepsilon]$ *and* $\pi_Q(\xi_x(t)) = \pi_Q(x')$ *for* $t \in]\varepsilon, \tau]$;
  2. $\pi_Q(x) = \pi_Q(x')$ *and* $\xi_x : \mathbb{R}^+_0 \to \mathbb{R}^n$ *is a solution of Σ satisfying* $\xi_x(\tau) = x'$ *and* $\pi_Q(\xi_x(t)) = \pi_Q(x)$ *for all* $t \in \mathbb{R}^+_0$;
- $Y = X/Q$;
- $H = \pi_Q$.

[2] We consider a curve ξ_x defined on the closed set $[0, \tau]$ while implicitly assuming the existence of a curve $\xi'_x :]a, b[\to \mathbb{R}^n$ satisfying $\xi_x = \xi'_x|_{[0,\tau]}$, and all the additional conditions imposed on ξ_x.

The construction of $S_{QL}(\Sigma)$ depends on Q to define both the outputs and the transitions. In fact, the transitions of $S_{QL}(\Sigma)$ were constructed to describe the observations that a user can make of Σ through the output map H. Instead of recording the outputs periodically, we record only the *changes* or *transitions* in the output signal. An output change, say from y to y', is witnessed by the transition $x \xrightarrow{\tau} \xi_x(\tau)$ of $S_{QL}(\Sigma)$ satisfying $H(x) = y$ and $H(x') = y'$. Note that we only allow $H \circ \xi$ to switch once from y to y'. The switching time is $\varepsilon \in [0, \tau]$ and two situations may occur: the point $\xi(\varepsilon)$ belongs to $H^{-1}(y')$, corresponding to 1a in Definition 7.2, or the point $\xi(\varepsilon)$ belongs to $H^{-1}(y)$, corresponding to 1b in Definition 7.2. This sampling strategy has two important consequences: the output strings never have repeated symbols, and trajectories ξ of unbounded duration may be sampled into finite strings. The last consequence occurs, *e.g.*, when $H \circ \xi(t) = H \circ \xi(0)$ for $t \in \mathbb{R}_0^+$. This is quite unfortunate since a finite observed string suggests that Σ is a blocking system even if this is not the case. To overcome this difficulty, we clairvoyantly repeat the last output y in an observed string induced by a trajectory ξ of unbounded duration. This is done by equipping $S_{QL}(\Sigma)$ with the transition $x \xrightarrow{\tau} \xi_x(\tau)$ when $H(x) = H \circ \xi_x(t)$ for $t \in \mathbb{R}_0^+$, as described in case 2 of Definition 7.2. To simplify notation, we denote $S_{QL}(\Sigma)$ by $S_Q(\Sigma)$ whenever $L = X$. A different way of modeling dynamical systems as systems, closer to the discussion in Section 1.3.2, appears in Chapter 10.

7.1.2 Hybrid dynamical systems

In addition to dynamical systems, we also consider hybrid dynamical systems consisting of a finite collection of dynamical systems and a set of rules describing when and how to switch among the individual dynamical systems. The definitions of hybrid dynamical system and its associated system are known to be difficult to digest. The reader in need of an aperitif is invited to reread Chapter 1 where hybrid systems are introduced through examples and several important notions such as invariant sets, guard sets, and reset functions are discussed.

Definition 7.3 (Hybrid dynamical system). *A hybrid dynamical system Σ is a quintuple $(S, \{\mathrm{In}_x\}_{x \in X}, \{\mathrm{Gu}_t\}_{t \in \longrightarrow}, \{\mathrm{Re}_t\}_{t \in \longrightarrow}, \{f_x\}_{x \in X})$ consisting of:*

- *a finite-state system $S = (X, U, \longrightarrow)$;*
- *a non-empty* invariant *set $\mathrm{In}_x \subseteq \mathbb{R}^n$ for each $x \in X$;*
- *a non-empty* guard *set $\mathrm{Gu}_{(x,u,x')} \subseteq \mathrm{In}_x$ for each $(x, u, x') \in \longrightarrow$;*
- *a* reset *function $\mathrm{Re}_{(x,u,x')} : \mathrm{In}_x \to \mathrm{In}_{x'}$ for each $(x, u, x') \in \longrightarrow$;*
- *a dynamical system (In_x, f_x) for each $x \in X$.*

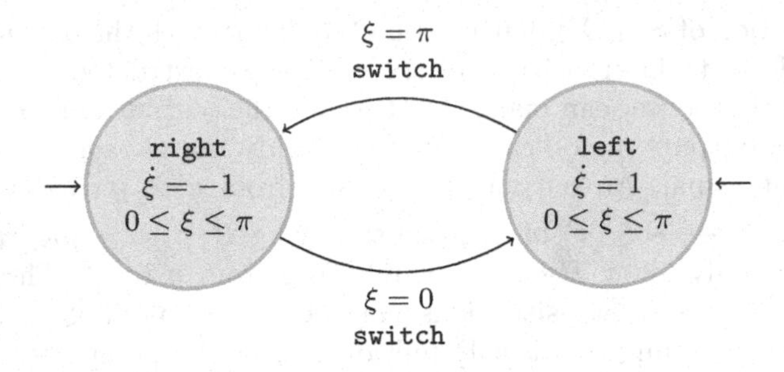

Fig. 7.1. Hybrid dynamical system modeling a windshield wiper.

As we distinguish between dynamical systems and control systems, we also distinguish between hybrid dynamical systems and hybrid control systems. The former are equipped with a dynamical system for each finite state while the later are equipped with a control system for each finite state.

Example 7.4. We consider a simplistic model for a windshield wiper given by the hybrid dynamical system Σ in Figure 7.1. The finite-state system S corresponding to Σ is defined by the set of states $X = \{\mathtt{right}, \mathtt{left}\}$, by the set of inputs $U = \{\mathtt{switch}\}$, and by the transition relation containing only two transitions $\mathtt{right} \xrightarrow{\mathtt{switch}} \mathtt{left}$ and $\mathtt{left} \xrightarrow{\mathtt{switch}} \mathtt{right}$. The finite state **right** describes the clockwise motion of the wiper while the finite state **left** describes its counterclockwise motion. The angle described by the wiper is denoted by ξ and taken as positive when measure in the counter clockwise direction. Its dynamics is given by the dynamical system $([0, \pi], -1)$ at the state **right** and by the dynamical system $([0, \pi], 1)$ at the state **left**. In particular, this implies $\mathrm{In}_{\mathtt{right}} = [0, \pi] = \mathrm{In}_{\mathtt{left}}$ since the wiper's angle changes between 0 and π radians. The analogue of a trajectory for a dynamical system is an execution for a hybrid dynamical system. Given an initial condition, *e.g.*, $(\mathtt{right}, \frac{\pi}{2}) \in X \times \mathrm{In}_{\mathtt{right}}$, the execution from $(\mathtt{right}, \frac{\pi}{2})$ starts with the solution $\xi_{\frac{\pi}{2}} = \frac{\pi}{2} - t$ of $([0, \pi], -1)$. At time $t = \frac{\pi}{2}$ we have $\xi_{\frac{\pi}{2}}(\frac{\pi}{2}) = \frac{\pi}{2} - \frac{\pi}{2} = 0$ and the guard associated with the transition $\mathtt{right} \xrightarrow{\mathtt{switch}} \mathtt{left}$, $\mathrm{Gu}_{(\mathtt{right},\mathtt{switch},\mathtt{left})} = \{0\}$, is satisfied, that is, $\xi_{\frac{\pi}{2}}(\frac{\pi}{2}) \in \mathrm{Gu}_{(\mathtt{right},\mathtt{switch},\mathtt{left})}$. The execution can be continued from $(\mathtt{right}, 0)$ by a transition from state **right** to state **left**. In this case, this transition has to be taken since failure to do so would force the angle to leave the invariant set. Taking the transition $\mathtt{right} \xrightarrow{\mathtt{switch}} \mathtt{left}$ changes the state from $(\mathtt{right}, 0)$ to $(\mathtt{left}, \mathrm{Re}_{(\mathtt{left},\mathtt{switch},\mathtt{right})}(0)) = (\mathtt{left}, 0)$. Upon taking the transition, the execution proceeds from $(\mathtt{left}, 0)$ with the evolution of the angle according to the solution of $([0, \pi], 1)$ until another transition is taken and so on. $\lhd$

The previous example suggested that executions of hybrid dynamical systems consist of sequences of continuous-time evolutions interleaved with transitions between finite states. We do not need to formally define executions since they arise as the internal behavior of the systems modeling hybrid dynamical systems. Therefore, system models for hybrid dynamical systems require transition relations describing both the continuous-time dynamics defined by the dynamical systems (In_x, f_x) as well as the dynamics induced by the finite-state system S through the guards and reset maps.

Definition 7.5. *Consider a hybrid dynamical system:*

$$\Sigma = (S_a, \{\mathrm{In}_{x_a}\}_{x_a \in X_a}, \{\mathrm{Gu}_{t_a}\}_{t_a \in \underset{a}{\longrightarrow}}, \{\mathrm{Re}_{t_a}\}_{t_a \in \underset{a}{\longrightarrow}}, \{f_{x_a}\}_{x_a \in X_a}).$$

For each $x_a \in X_a$ let Q_{x_a} be a finite equivalence relation on In_{x_a}, and let $L \subseteq \{(x_a, x_b) \in X_a \times \mathbb{R}^n \mid x_b \in \mathrm{In}_{x_a}\}$ be a set of initial states. The system associated with Σ, $\{Q_{x_a}\}_{x_a \in X_a}$, and L, denoted by $S_{QL}(\Sigma)$, consists of:

- $X = \{(x_a, x_b) \in X_a \times \mathbb{R}^n \mid x_b \in \mathrm{In}_{x_a}\}$;
- $X_0 = L$;
- $U = U_a \cup \mathbb{R}_0^+$;
- $(x_a, x_b) \overset{u}{\longrightarrow} (x'_a, x'_b)$ *if one of the following three conditions holds:*
  1. $u \in U_a$, $x_b \in \mathrm{Gu}_{(x_a,u,x'_a)}$, *and* $x'_b = \mathrm{Re}_{(x_a,u,x'_a)}(x_b)$;
  2. $u \in \mathbb{R}_0^+$, $x'_a = x_a$, $\pi_{Q_{x_a}}(x_b) \neq \pi_{Q_{x_a}}(x'_b)$, $\xi_{x_b} : [0, u] \to \mathrm{In}_{x_a}$ *is a solution of* $(\mathrm{In}_{x_a}, f_{x_a})$ *satisfying* $\xi_{x_b}(u) = x'_b$, *and there exists* $\varepsilon \in [0, u]$ *satisfying one of the following:*
     a) $\pi_{Q_{x_a}}(\xi_{x_b}(t)) = \pi_{Q_{x_a}}(x_b)$ *for* $t \in [0, \varepsilon[$ *and* $\pi_{Q_{x_a}}(\xi_{x_b}(t)) = \pi_{Q_{x_a}}(x'_b)$ *for* $t \in [\varepsilon, u]$;
     b) $\pi_{Q_{x_a}}(\xi_{x_b}(t)) = \pi_{Q_{x_a}}(x_b)$ *for* $t \in [0, \varepsilon]$ *and* $\pi_{Q_{x_a}}(\xi_{x_b}(t)) = \pi_{Q_{x_a}}(x'_b)$ *for* $t \in]\varepsilon, u]$;
  3. $u \in \mathbb{R}_0^+$, $x'_a = x_a$, $\pi_{Q_{x_a}}(x_b) = \pi_{Q_{x_a}}(x'_b)$, $\xi_{x_b} : \mathbb{R}_0^+ \to \mathrm{In}_{x_a}$ *is a solution of* $(\mathrm{In}_{x_a}, f_{x_a})$ *satisfying* $\xi_{x_b}(u) = x'_b$, *and* $\pi_{Q_{x_a}}(\xi_{x_b}(t)) = \pi_{Q_{x_a}}(x_b)$ *for all* $t \in \mathbb{R}_0^+$.
- $Y = \left\{(x_a, y) \in X_a \times \bigcup_{x'_a \in X_a} \mathrm{In}_{x'_a}/Q_{x'_a} \mid y \in \mathrm{In}_{x_a}/Q_{x_a}\right\}$;
- $H(x_a, x_b) = (x_a, \pi_{Q_{x_a}}(x_b))$.

For a hybrid dynamical system Σ, the state of the system $S_{QL}(\Sigma)$ is pair $(x_a, x_b) \in X_a \times \mathrm{In}_{x_a}$ consisting of a finite part x_a, inherited from the finite-state system S_a, and an infinite part x_b, inherited from the dynamical system $(\mathrm{In}_{x_a}, f_{x_a})$. The evolution of the finite part is governed by a coupling between the transition relation of S_a and the transition relation of $S_{Q_{x_a}}(\mathrm{In}_{x_a}, f_{x_a})$. A transition $x_a \underset{a}{\overset{u_a}{\longrightarrow}} x'_a$ in S_a defines a transition $(x_a, x_b) \overset{u_a}{\longrightarrow} (x'_a, x'_b)$ in $S_{QL}(\Sigma)$ only when the infinite part of the state belongs to the guard $\mathrm{Gu}_{(x_a,u_a,x'_a)}$ and when x'_b is obtained from x_b by applying the reset map $\mathrm{Re}_{(x_a,u_a,x'_a)}$. This kind of transitions, corresponding to case 1 in Definition 7.5, are termed discrete transitions since they are induced by the transitions in S_a. In addition to discrete transitions there are also continuous flows described by cases 2 and 3. A transition $(x_a, x_b) \xrightarrow{u} (x'_a, x'_b)$ describes a continuous flow when $u \in \mathbb{R}_0^+$ is the duration of the flow defined by the solution ξ_{x_b} of $(\text{In}_{x_a}, f_{x_a})$ satisfying $\xi_{x_b}(u) = x'_b$. Note that a continuous flow leaves the finite part of the state unaltered, *i.e.*, $x'_a = x_a$. Consequently, during a continuous flow, a hybrid dynamical system behaves as the dynamical system $(\text{In}_{x_a}, f_{x_a})$ and the construction of $S_{QL}(\Sigma)$ mirrors the construction of $S_{Q_{x_a}}(\text{In}_{x_a}, f_{x_a})$ in Definition 7.2. The construction in Definition 7.5 is the starting point for the different abstraction techniques described in this chapter.

7.2 Timed automata

Several abstraction techniques for hybrid systems are generalizations of a construction developed for a special class of hybrid dynamical systems: timed automata. This class was originally introduced to reason about the temporal properties of software systems and is characterized by restrictions on the invariants, guards, resets, and continuous-time dynamics.

Definition 7.6 (Timed automaton). *A hybrid dynamical system:*

$$\Sigma = (S_a, \{\text{In}_{x_a}\}_{x_a \in X_a}, \{\text{Gu}_{t_a}\}_{t_a \in \underset{a}{\longrightarrow}}, \{\text{Re}_{t_a}\}_{t_a \in \underset{a}{\longrightarrow}}, \{f_{x_a}\}_{x_a \in X_a}).$$

is said to be a timed automaton *if the following four conditions are satisfied:*

1. *for every $x_a \in X_a$, the sets $\text{In}_{x_a} \subseteq \mathbb{R}^n$ are defined by finitely many conjunctions of conditions of the form $x_{bi} \sim c$ with $i \in \{1, 2, \ldots, n\}$, $\sim \in \{\leq, <, =, >, \geq\}$, and $c \in \mathbb{Q}_0^+$;*
2. *for every $(x_a, u_a, x'_a) \in \underset{a}{\longrightarrow}$, the sets $\text{Gu}_{(x_a, u_a, x'_a)}$ are defined by finitely many conjunctions of conditions of the form $x_{bi} \sim c$ with $i \in \{1, 2, \ldots, n\}$, $\sim \in \{\leq, <, =, >, \geq\}$, and $c \in \mathbb{Q}_0^+$;*
3. *for every $(x_a, u_a, x'_a) \in \underset{a}{\longrightarrow}$, $x_b \in \text{In}_{x_a}$, and $i \in \{1, 2, \ldots, n\}$, the composite $\pi_i \circ \text{Re}_{(x_a, u_a, x'_a)}(x_b)$ is either x_{bi} or 0;*
4. *for every $x_a \in X_a$, $x_b \in \text{In}_{x_a}$, and $i \in \{1, 2, \ldots, n\}$, $\pi_i \circ f_{x_a}(x_b) = 1$.*

We explain the meaning of the preceding restrictions by referring to the timed automaton, modeling a periodic real-time task, presented in Chapter 1, Figure 1.6. For convenience, this timed automaton is also depicted in Figure 7.2. The invariant associated with any of the finite states is a conjunction of inequalities in which an infinite state variable x_{bi} is compared with a non-negative constant. The invariant associated with the state `active` is $0 \leq x_{b1} \leq D$, the conjunction of $x_{b1} \geq 0$ with $x_{b1} \leq D$. The guards are defined by sets constructed in the same fashion. The guard associated with the transition `active` $\xrightarrow{\texttt{expired}}$ `error` is $x_{b1} = D$ and of the form required by the definition of timed automata. Consider now the transition

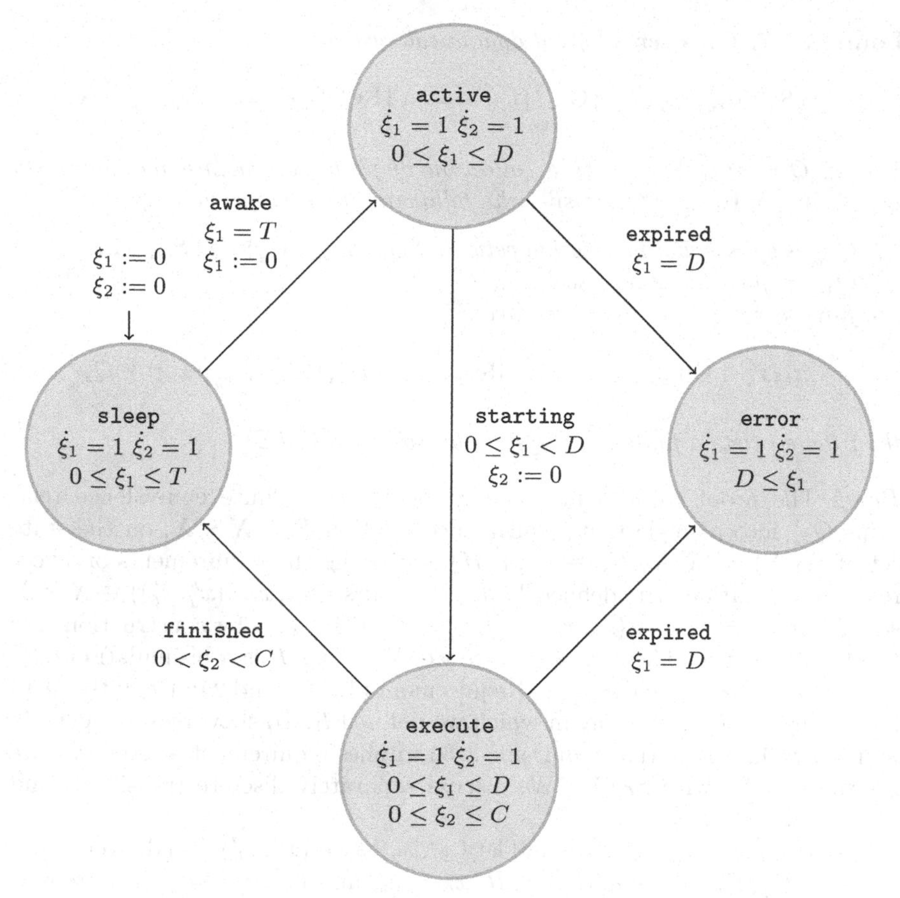

Fig. 7.2. Timed automaton representing a periodic real-time task.

$\texttt{active} \xrightarrow{\texttt{starting}} \texttt{execute}$ which is decorated with $x_{b2} := 0$. The assignment $x_{b2} := 0$ defines the reset map associated with this transition to be $\text{Re}_{(\texttt{active},\texttt{starting},\texttt{execute})}(x_b) = (x_{b1}, 0)$ by stating that x_{b2} should be reset to zero while x_{b1} should remain unchanged. Finally, the differential equations in each of the finite states are of the form $\dot{\xi}_1 = 1$ and $\dot{\xi}_2 = 1$.

Existence of a finite-state system bisimilar to the infinite-state system S associated with a timed automaton implies that for each finite state $x_a \in X_a$ of S, there exists a finite-state system bisimulating the system $S_Q(\text{In}_{x_a}, f_{x_a})$ for some finite equivalence relation $Q \subseteq \text{In}_{x_a} \times \text{In}_{x_a}$. The converse of this observation does not hold unless we make some additional assumptions. The next result, a direct consequence of Theorem 4.18, offers the missing assumptions in the general context of hybrid systems.

Lemma 7.7. *Consider a hybrid dynamical system:*

$$\Sigma = (S_a, \{\text{In}_{x_a}\}_{x_a \in X_a}, \{\text{Gu}_{t_a}\}_{t_a \in \xrightarrow[a]{}}, \{\text{Re}_{t_a}\}_{t_a \in \xrightarrow[a]{}}, \{f_{x_a}\}_{x_a \in X_a})$$

and let $Q = \{Q_{x_a}\}_{x_a \in X_a}$ be a collection of finite equivalence relations with $Q_{x_a} \subseteq \text{In}_{x_a} \times \text{In}_{x_a}$. If Q satisfies the following three properties:

1. *Q_{x_a} is a bisimulation relation between $S_{Q_{x_a}}(\text{In}_{x_a}, f_{x_a})$ and $S_{Q_{x_a}}(\text{In}_{x_a}, f_{x_a})$;*
2. *Q_{x_a} respects the guard sets;*
3. *for every $(x_b, x_b') \in Q_{x_a}$ we have:*

$$x_b, x_b' \in \text{Gu}_{(x_a,u,x_a')} \implies (\text{Re}_{(x_a,u,x_a')}(x_b), \text{Re}_{(x_a,u,x_a')}(x_b')) \in Q_{x_a'},$$

then there exists a finite-state system bisimilar to $S_Q(\Sigma)$.

Proof. The proof consists in weaving together the finite equivalence relations Q_{x_a} into a single finite equivalence relation $R \subseteq X \times X$, on the state set of $S_Q(\Sigma) = (X, X_0, U, \longrightarrow, Y, H)$, satisfying the requirements of Theorem 4.18. Relation R is defined by all the pairs $((x_a, x_b), (x_a', x_b')) \in X \times X$ satisfying $x_a = x_a'$ and $(x_b, x_b') \in Q_{x_a}$. Since R is finite by construction, the result follows from Theorem 4.18 once we show that R is a bisimulation relation between $S_Q(\Sigma)$ and $S_Q(\Sigma)$. Requirements 1a, 1b, and 2 in Definition 4.13 follow directly from the way in which we defined R. To show requirements 3a and 3b, we first note that requirement 3a implies requirement 3b since we are relating $S_Q(\Sigma)$ with $S_Q(\Sigma)$. We discuss separately discrete transitions and continuous flows.

Let $((x_a, x_b), (x_a', x_b')) \in R$ and let $(x_a, x_b) \xrightarrow{u} (x_a'', x_b'')$ be a discrete transition in $S_Q(\Sigma)$. By definition of R, $x_a = x_a'$ and $(x_b, x_b') \in Q_{x_a}$. Moreover, by definition of $S_Q(\Sigma)$, $x_b \in \text{Gu}_{(x_a,u,x_a'')}$. Since Q_{x_a} respects guard sets, x_b' also belongs to $\text{Gu}_{(x_a,u,x_a'')}$. It now follows from the third property of Q that $(\text{Re}_{(x_a,u,x_a')}(x_b), \text{Re}_{(x_a,u,x_a')}(x_b')) = (x_b'', x_b''') \in Q_{x_a''}$. Consequently, by construction of $S_Q(\Sigma)$, we have the transition $(x_a', x_b') \xrightarrow{u} (x_a'', x_b''')$ in $S_Q(\Sigma)$. Since $(x_b'', x_b''') \in Q_{x_a''}$ implies $((x_a'', x_b''), (x_a'', x_b''')) \in R$ we conclude that requirement 3a in Definition 4.13 holds for discrete transitions.

Consider now a continuous flow $(x_a, x_b) \xrightarrow{u} (x_a'', x_b'')$ in $S_Q(\Sigma)$ and let $((x_a, x_b), (x_a', x_b')) \in R$, *i.e.*, $x_a = x_a'$ and $(x_b, x_b') \in Q_{x_a}$. By definition of continuous flow, $x_a = x_a''$ and $x_b \xrightarrow{u} x_b''$ is a transition in $S_{Q_{x_a}}(\text{In}_{x_a}, f_{x_a})$. Since Q_{x_a} is a bisimulation relation between $S_{Q_{x_a}}(\text{In}_{x_a}, f_{x_a})$ and $S_{Q_{x_a}}(\text{In}_{x_a}, f_{x_a})$, there exists a transition $x_b' \xrightarrow{u'} x_b'''$ in $S_{Q_{x_a}}(\text{In}_{x_a}, f_{x_a})$ satisfying $(x_b'', x_b''') \in Q_{x_a}$. We thus conclude the existence of the continuous flow $(x_a, x_b') \xrightarrow{u'} (x_a, x_b''')$ in $S_Q(\Sigma)$ satisfying $((x_a, x_b'), (x_a, x_b''')) \in R$ which concludes the proof. □

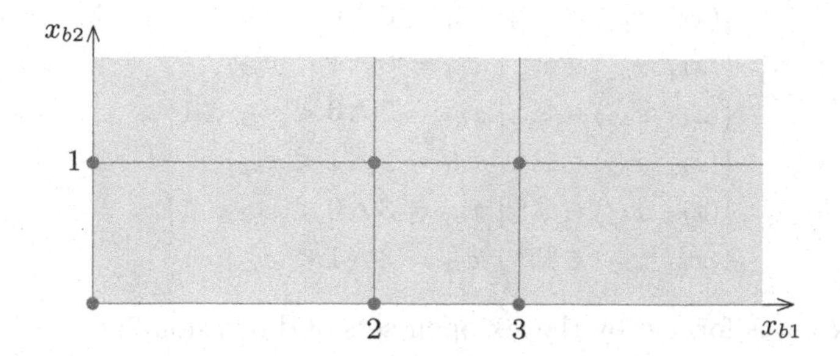

Fig. 7.3. Partition of $(\mathbb{R}_0^+)^2$, induced by the equivalence relation Q, distinguishing guards, and invariant sets.

Lemma 7.7 can be straightforwardly applied to hybrid dynamical systems with constant reset maps since in such cases condition 3 is automatically satisfied in virtue of the equality $\mathrm{Re}_{(x_a,u,x'_a)}(x_b) = \mathrm{Re}_{(x_a,u,x'_a)}(x'_b)$. Hence, it suffices to construct finite-state abstractions for the systems $S_{Q_{x_a}}(\mathrm{In}_{x_a}, f_{x_a})$, based on equivalence relations Q_{x_a} respecting the guard sets, in order to obtain a finite-state abstraction for $S_Q(\Sigma)$. This observation justifies why we devote a large portion of this chapter to the construction of finite-state abstractions of several classes of dynamical systems.

For timed automata, we do not need to rely on constant reset maps due to the simple nature of the differential equations describing the continuous-time dynamics. We illustrate the construction of a family of equivalence relations $\{Q_{x_a}\}_{x_a \in X_a}$ satisfying the conditions in Lemma 7.7 using again the timed automaton in Figure 7.2. For concreteness, we take $C = 1$, $D = 2$, and $T = 3$. Instead of constructing directly an equivalence relation for each invariant set, we start by constructing a single equivalence relation Q on $(\mathbb{R}_0^+)^2$ that respects all the invariant sets and all the guards. Relation Q is displayed in Figure 7.3 and defines three different kinds of equivalence classes. The first kind, of dimension zero, consists of six singleton sets:

$$\{(0,0)\}, \quad \{(2,0)\}, \quad \{(3,0)\}, \quad \{(0,1)\}, \quad \{(2,1)\}, \quad \{(3,1)\}.$$

The second kind, of dimension 1, consists of twelve horizontal and vertical open line segments:

$$\begin{aligned}
&\{(x_{b1}, x_{b2}) \in \mathbb{R}^2 \mid 0 < x_{b1} < 2 \wedge x_{b2} = 0\},\\
&\{(x_{b1}, x_{b2}) \in \mathbb{R}^2 \mid 2 < x_{b1} < 3 \wedge x_{b2} = 0\},\\
&\{(x_{b1}, x_{b2}) \in \mathbb{R}^2 \mid 3 < x_{b1} \wedge x_{b2} = 0\},\\
&\{(x_{b1}, x_{b2}) \in \mathbb{R}^2 \mid 0 < x_{b1} < 2 \wedge x_{b2} = 1\},\\
&\{(x_{b1}, x_{b2}) \in \mathbb{R}^2 \mid 2 < x_{b1} < 3 \wedge x_{b2} = 1\},\\
&\{(x_{b1}, x_{b2}) \in \mathbb{R}^2 \mid 3 < x_{b1} \wedge x_{b2} = 1\},
\end{aligned}$$

$$\{(x_{b1}, x_{b2}) \in \mathbb{R}^2 \mid x_{b1} = 0 \wedge 0 < x_{b2} < 1\},$$
$$\{(x_{b1}, x_{b2}) \in \mathbb{R}^2 \mid x_{b1} = 0 \wedge 1 < x_{b2}\},$$
$$\{(x_{b1}, x_{b2}) \in \mathbb{R}^2 \mid x_{b1} = 2 \wedge 0 < x_{b2} < 1\},$$
$$\{(x_{b1}, x_{b2}) \in \mathbb{R}^2 \mid x_{b1} = 2 \wedge 1 < x_{b2}\},$$
$$\{(x_{b1}, x_{b2}) \in \mathbb{R}^2 \mid x_{b1} = 3 \wedge 0 < x_{b2} < 1\},$$
$$\{(x_{b1}, x_{b2}) \in \mathbb{R}^2 \mid x_{b1} = 3 \wedge 1 < x_{b2}\}.$$

The third kind, is formed by the six open sets of dimension 2:

$$\{(x_{b1}, x_{b2}) \in \mathbb{R}^2 \mid 0 < x_{b1} < 2 \wedge 0 < x_{b2} < 1\},$$
$$\{(x_{b1}, x_{b2}) \in \mathbb{R}^2 \mid 2 < x_{b1} < 3 \wedge 0 < x_{b2} < 1\},$$
$$\{(x_{b1}, x_{b2}) \in \mathbb{R}^2 \mid 3 < x_{b1} \wedge 0 < x_{b2} < 1\},$$
$$\{(x_{b1}, x_{b2}) \in \mathbb{R}^2 \mid 0 < x_{b1} < 2 \wedge 1 < x_{b2}\},$$
$$\{(x_{b1}, x_{b2}) \in \mathbb{R}^2 \mid 2 < x_{b1} < 3 \wedge 1 < x_{b2}\},$$
$$\{(x_{b1}, x_{b2}) \in \mathbb{R}^2 \mid 3 < x_{b1} \wedge 1 < x_{b2}\}.$$

Although the equivalence relation Q respects invariant sets and guards, it fails to respect the continuous flow in the sense that different points in the same equivalence class may visit, under the continuous flow, different equivalence classes. This is illustrated in Figure 7.4 where the flow is represented by the dashed lines. Since failure to respect the continuous flow implies failure of the first assumption in Lemma 7.7, we refine Q to an equivalence relation Q' respecting the continuous flow. Given the special nature of the differential equations present in timed automata, the continuous flow with initial condition (x_{b1}, x_{b2}) is the straight line with unit slope passing through (x_{b1}, x_{b2}). Using this fact, it is straightforward to refine Q to the equivalence relation Q' represented in Figure 7.5. The reader should verify that any two points in the same equivalence class of Q' are taken by the continuous flow through the same equivalence classes. Upon inspection of Figure 7.5, we observe that the equivalence classes of Q' are not closed under projection maps. To be

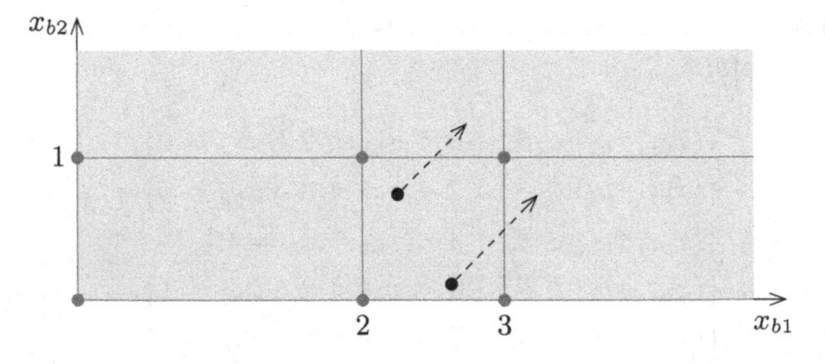

Fig. 7.4. Two points on the same equivalence taken by the continuous flow to different equivalence classes.

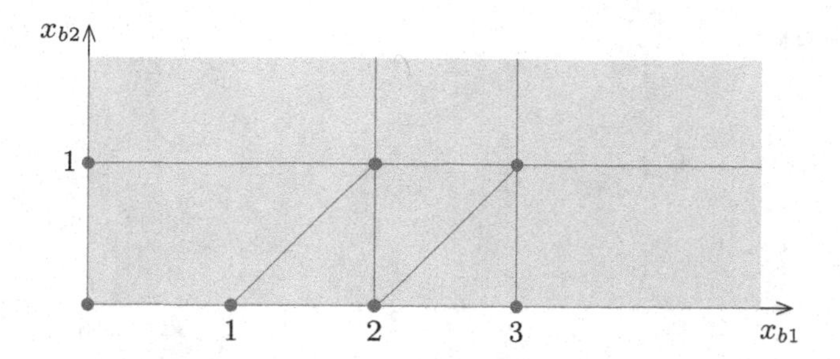

Fig. 7.5. Partition of $(\mathbb{R}_0^+)^2$, induced by equivalence relation Q' refining equivalence relation Q. Guard and invariant sets are distinguished while the continuous flow is respected.

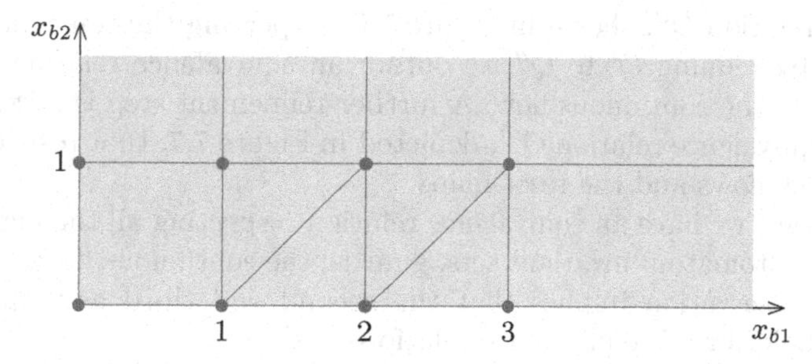

Fig. 7.6. Partition of $(\mathbb{R}_0^+)^2$, induced by equivalence relation Q'' refining equivalence relation Q'. Guard and invariant sets are distinguished while reset maps are respected.

precise, consider the maps $r_1 : (\mathbb{R}_0^+)^2 \to (\mathbb{R}_0^+)^2$, $r_2 : (\mathbb{R}_0^+)^2 \to (\mathbb{R}_0^+)^2$, and $r_{12} : (\mathbb{R}_0^+)^2 \to (\mathbb{R}_0^+)^2$ defined by:

$$r_1(x_{b1}, x_{b2}) = (x_{b1}, 0), \quad r_2(x_{b1}, x_{b2}) = (0, x_{b2}), \quad r_{12}(x_{b1}, x_{b2}) = (0, 0).$$

For any equivalence class $[x_b] \in (\mathbb{R}_0^+)^2/Q'$ and for any map $g \in \{1_{(\mathbb{R}_0^+)^2}, r_2, r_{12}\}$, $g([x_b])$ is an equivalence class of Q'. However, this is no longer the case when $g = r_1$. In particular, applying r_1 to the equivalence class defined by:

$$\{(x_{b1}, x_{b_2}) \in (\mathbb{R}_0^+)^2 \mid 0 < x_{b1} < 2 \wedge x_{b2} = 1\}$$

results in the set:

$$\{(x_{b1}, x_{b_2}) \in (\mathbb{R}_0^+)^2 \mid 0 < x_{b1} < 2 \wedge x_{b2} = 0\}$$

which clearly does not belong to $(\mathbb{R}_0^+)^2/Q'$. Since the reset maps are necessarily of the form $1_{(\mathbb{R}_0^+)^2}$, r_1, r_2, and r_{12}, we conclude that Q' does not respect the reset maps. Failure to respect the reset maps implies failure of the third assumption in Lemma 7.7. The next step is then obvious: refine Q' to the

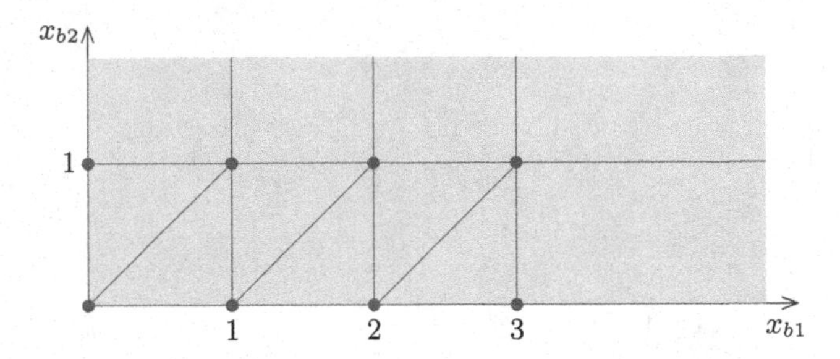

Fig. 7.7. Partition of $(\mathbb{R}_0^+)^2$, induced by equivalence relation Q''' refining equivalence relation Q''. Guard and invariant sets are distinguished while reset maps and the continuous flow are respected.

equivalence relation Q'', shown in Figure 7.6, respecting the reset maps. Unfortunately, by refining Q' to Q'' we obtain an equivalence relation that no longer respects the continuous flow. A further refinement step is necessary to obtain the equivalence relation Q''', depicted in Figure 7.7, that respects both the continuous flows and the reset maps.

At this point we have an equivalence relation respecting all the ingredients of our timed automaton: invariant sets, guards, the continuous flow, and reset maps. This observation implies that the second and third assumptions in Lemma 7.7 hold for the equivalence relations:

$$Q_{x_a} = (\text{In}_{x_a} \times \text{In}_{x_a}) \cap Q'', \quad x_a \in X_a.$$

Furthermore, as these equivalence relations also respect the continuous flow, they are bisimulation relations between $S_{Q_{x_a}}(\text{In}_{x_a}, f_{x_a})$ and $S_{Q_{x_a}}(\text{In}_{x_a}, f_{x_a})$. Lemma 7.7 then guarantees the existence of a finite-state system bisimilar to $S_Q(\Sigma)$.

The previously described steps lead to the following general result.

Theorem 7.8. *Let Σ be a timed automaton and let $Q = \{Q_{x_a}\}_{x_a \in X_a}$ be a collection of finite equivalence relations $Q_{x_a} \subseteq \text{In}_{x_a} \times \text{In}_{x_a}$ whose equivalence classes are defined by finitely many conjunctions of conditions of the form $x_{bi} \sim c$ with $i \in \{1, 2, \ldots, n\}$, $\sim \in \{\leq, <, =, >, \geq\}$, and $c \in \mathbb{Q}_0^+$. Then, there exists a finite-state system bisimilar to $S_Q(\Sigma)$.*

The strategy to obtain Theorem 7.8 consisted in refining an initial finite equivalence relation until invariants, guards, resets, and the continuous flow were respected. The resulting finite equivalence relation lead directly to the quotient system which is finite-state and bisimilar to the original system. We refer to these abstractions as quotient based abstractions. The existence of quotient based abstractions for timed automata can be further exploited to prove several deep and beautiful results for this class of systems. Instead of exploring further timed automata, we veer to more general classes of hybrid dynamical systems possessing richer continuous-time dynamics.

7.3 Order minimal hybrid dynamical systems

Polynomial functions have remarkable finiteness properties that are highly desirable for the construction of finite-state models for verification. Unfortunately, even if a differential equation is linear, its solutions are not, in general, polynomial functions of time. We thus need a larger class of sets and functions that still enjoys the same finiteness properties of polynomials and contains the solutions of (some) differential equations. Order minimal structures provide an axiomatization for a class that is general enough to include semi-linear, semi-algebraic, and even more complicated sets and functions, while still possessing very strong finiteness properties. In what follows, we restrict the discussion to order minimal structures over the real numbers.

Definition 7.9 (Order minimal structure). *An* order minimal structure *on* $\mathbb{R}$ *is a sequence* $\mathcal{S} = \{\mathcal{S}_n\}_{n\in\mathbb{N}}$ *such that for each* $n \in \mathbb{N}$ *we have:*

1. $\mathcal{S}_n$ *is a boolean algebra of subsets of* $\mathbb{R}^n$*, i.e.,* $\mathcal{S}_n$ *is a collection of subsets of* $\mathbb{R}^n$ *such that:*
   a) $\varnothing \in \mathcal{S}_n$;
   b) $Z, W \in \mathcal{S}_n$ *implies* $Z \cup W \in \mathcal{S}_n$;
   c) $Z \in \mathcal{S}_n$ *implies* $\mathbb{R}^n \backslash Z \in \mathcal{S}_n$;
2. $Z \in \mathcal{S}_n$ *implies* $Z \times \mathbb{R} \in \mathcal{S}_{n+1}$ *and* $\mathbb{R} \times Z \in \mathcal{S}_{n+1}$;
3. $\{(x_1, x_2, \ldots, x_n) \in \mathbb{R}^n \mid x_i = x_j\} \in \mathcal{S}_n$ *for* $1 \leq i < j \leq n$;
4. $Z \in \mathcal{S}_{n+1}$ *implies* $\pi(Z) \in \mathcal{S}_n$ *where* $\pi : \mathbb{R}^{n+1} \to \mathbb{R}^n$ *is the usual projection taking* $(x_1, \ldots, x_{n-1}, x_n)$ *to* $(x_1, \ldots, x_{n-1})$;
5. $\{r\} \in \mathcal{S}_1$ *for each* $r \in \mathbb{R}$ *and* $\{(x_1, x_2) \in \mathbb{R}^2 \mid x_1 < x_2\} \in \mathcal{S}_2$;
6. *the only sets in* $\mathcal{S}_1$ *are the finite unions of open intervals and points.*

Example 7.10. The class of *semi-linear sets* on $\mathbb{R}^n$ is defined by finite unions of subsets of $\mathbb{R}^n$ satisfying conjunctions of conditions of the form $f \sim 0$ where f is an affine function and $\sim \in \{=, >\}$. Semi-linear sets form an order minimal structure. When the functions f are polynomial instead of affine, we obtain the class of *semi-algebraic sets*. This class constitutes another example of an order minimal structure.

The set W, depicted in Figure 7.8, is a semi-algebraic set defined by:

$$W = \{(x_1, x_2) \in \mathbb{R}^2 \mid x_1^2 + x_2^2 \leq 1 \wedge -x_2 + x_1 \leq 0\} \tag{7.4}$$
$$\cup \{(x_1, x_2) \in \mathbb{R}^2 \mid x_1^2 + x_2^2 \leq 1 \wedge x_2 + x_1 \leq 0\}. \tag{7.5}$$

Although the sets (7.4) and (7.5) are not defined by conditions of the form $f \sim 0$ with $\sim \in \{=, >\}$, they can be rewritten in that form. For example (7.4) can be written as:

$$\{(x_1, x_2) \in \mathbb{R}^2 \mid -x_1^2 - x_2^2 + 1 > 0 \wedge x_2 - x_1 > 0\} \tag{7.6}$$
$$\cup \{(x_1, x_2) \in \mathbb{R}^2 \mid -x_1^2 - x_2^2 + 1 = 0 \wedge x_2 - x_1 = 0\}. \tag{7.7}$$

◁

Fig. 7.8. Semi-algebraic set defined by (7.4) and (7.5).

Given an order minimal structure $\mathcal{S}$ on $\mathbb{R}$, we say that a set $Z \subseteq \mathbb{R}^m$ is *definable* (in the structure $\mathcal{S}$) if $Z \in \mathcal{S}_m$. A map $f : Z \to \mathbb{R}^n$ is said to be definable (in the structure $\mathcal{S}$) if its graph $\Gamma(f) \subseteq \mathbb{R}^m \times \mathbb{R}^n$ is definable, *i.e.*, $\Gamma(f) \in \mathcal{S}_{n+m}$.

Let $W \subseteq \mathbb{R}^m \times \mathbb{R}^n$ be a definable set. For every $z \in \mathbb{R}^m$ the following set is a well defined subset of $\mathbb{R}^n$ called the *fiber* of W over z:

$$W_z = \{z' \in \mathbb{R}^n \mid (z, z') \in W\}.$$

We can regard W as a family of sets in $\mathbb{R}^n$ parameterized by points $z \in \mathbb{R}^m$. The only result we need from the theory of order minimal structures shows the existence of an upper bound for the number of connected components of W_z which is independent of z.

Theorem 7.11 (Uniform finiteness). *Let $W \subseteq \mathbb{R}^m \times \mathbb{R}^n$ be a definable set. Then, there is a number $k \in \mathbb{N}$ such that for every $z \in \mathbb{R}^m$, the fiber W_z has at most k connected components.*

Example 7.12. Consider the set $W \subseteq \mathbb{R} \times \mathbb{R}$ in Figure 7.8. For $1 < z < -1$, $W_z = \varnothing$; for $-1 \leq z \leq 0$, W_z consists of a single connected component; and for $0 < z \leq 1$, W_z consists of two connected components. Although the number of connected components changes with z, Theorem 7.11 guarantees the existence of a uniform upper-bound which is this case is 2. $\lhd$

In addition to semi-linear and semi-algebraic sets there are two other order minimal structures that will play an important role in this section. The first consists of *semi-exponential-algebraic sets*. Such sets are finite unions of sets in $\mathbb{R}^n$ described by conjunctions of conditions of the form $f \sim 0$ where the functions f are polynomials in the variables $x_1, x_2, \ldots, x_n, e^{x_1}, e^{x_2}, \ldots, e^{x_n}$ and $\sim \in \{=, >\}$. The fact that we can work with the exponential function will be essential when dealing with solutions of linear differential equations. The second structure extends semi-exponential-algebraic sets with any finite number of analytic functions $h : \mathbb{R}^n \to \mathbb{R}$ restricted to the hyper-cube $[-1, 1]^n$. This means that the functions f are polynomials in the variables $x_1, x_2, \ldots, x_n, e^{x_1}, e^{x_2}, \ldots, e^{x_n}, h_1(x), h_2(x), \ldots, h_k(x)$ for analytic functions $h_i : \mathbb{R}^n \to \mathbb{R}$ restricted to $[-1, 1]^n$.

We now consider dynamical systems $\Sigma = (\mathbb{R}^n, \theta)$ in which $\theta : \mathbb{R}^n \times \mathbb{R} \to \mathbb{R}^n$ is a definable map. The finiteness properties of θ, ensured by definability, can be used to guarantee the existence of finite-state abstractions. The starting point is a dynamical system $\Sigma = (\mathbb{R}^n, \theta)$ and an equivalence relation Q on $\mathbb{R}^n$ with finitely many equivalence classes consisting of definable sets. Such equivalence relations are termed definable finite equivalence relations.

Theorem 7.13. *Let $\Sigma = (\mathbb{R}^n, \theta)$ be a complete dynamical system in which $\theta : \mathbb{R}^n \times \mathbb{R} \to \mathbb{R}^n$ is a definable map. For any definable finite equivalence relation Q on $\mathbb{R}^n$, there exists a finite-state system bisimilar to $S_Q(\Sigma)$.*

Proof. Recall that since Σ is a complete dynamical system, for every $x \in \mathbb{R}^n$ there is a unique solution $\xi_x : \mathbb{R} \to \mathbb{R}^n$ satisfying $\xi_x(0) = x$. Let $\mathcal{P}$ be the partition of $\mathbb{R}^n$ induced by Q. Then $\xi_x^{-1}(\mathcal{P})$ is a partition of $\mathbb{R}$ that we denote by $\mathcal{P}_{\mathbb{R}}(x)$. The cardinality of $\mathcal{P}_{\mathbb{R}}(x)$ corresponds to the number (with multiplicity) of equivalence classes visited (in positive and negative time) by the solution ξ_x. We now use Theorem 7.11 to assert the existence of a global bound for the cardinality of $\mathcal{P}_{\mathbb{R}}(x)$ independent of $x \in \mathbb{R}^n$. To do this, we let $Z(q)$ be the set $\theta^{-1}(q)$ where q is an equivalence class of Q, thus $(x,t) \in Z(q)$ if $\theta(x,t) \in q$. By Theorem 7.11, there is a uniform upper-bound on the cardinality of the fibers $Z_x(q) = \{t \in \mathbb{R} \mid \theta(x,t) \in q\}$. This upper-bound describes how many times the equivalence class q is visited by ξ_x. Moreover, since there are finitely many equivalence classes q, there is a finite upper-bound for the cardinality of $\theta^{-1}(\mathcal{P})$ and therefore also for the cardinality of $\mathcal{P}_{\mathbb{R}}(x)$, independently of $x \in \mathbb{R}^n$. To each point $x \in \mathbb{R}^n$ we can now associate a finite string $\mathsf{q}(x)$ obtained by concatenating elements in $\{q_1, \ldots, q_k\} \cup \{\widehat{q}_1, \ldots, \widehat{q}_k\}$ where each q_i is an equivalence class of Q. This string describes the equivalence classes of Q visited along ξ_x, and is of the form $\mathsf{q}(x) = q_0 q_1 \ldots q_{i-1} \widehat{q}_i q_{i+1} \ldots q_l$ with q_j representing the equivalence classes of Q to be visited in the future when $j > i$, visited in the past when $j < i$, and $\widehat{q}_i$ representing the equivalence class containing x. The set of all such strings is finite, since there is a uniform upper bound on the cardinality of $\mathcal{P}_{\mathbb{R}}(x)$, and defines the state set X of the finite-state system S. The remaining elements of $S = (X, U, \longrightarrow, Y, H)$ are as follows:

$$U = \{*\}, \quad Y = X/Q, \quad H(q_0 q_1 \ldots q_{i-1} \widehat{q}_i q_{i+1} \ldots q_l) = q_i$$

with $\longrightarrow$ consisting of two kinds of transitions:

1. $q_0 q_1 \ldots q_{i-1} \widehat{q}_i q_{i+1} \ldots q_l \xrightarrow{*} q_0 q_1 \ldots q_{i-1} q_i \widehat{q}_{i+1} \ldots q_l$;
2. $q_0 q_1 \ldots q_{l-1} \widehat{q}_l \xrightarrow{*} q_0 q_1 \ldots q_{l-1} \widehat{q}_l$.

It is now routine to check that the relation $R \subseteq \mathbb{R}^n \times X$ defined by all the pairs $(x, \mathsf{q}(x))$ is a bisimulation relation between $S_Q(\Sigma)$ and S. □

The proof of Theorem 7.13 can be seen as an application of Theorem 4.18 to the system $S_Q(\Sigma)$ and to the equivalence relation $R \subseteq \mathbb{R}^n \times \mathbb{R}^n$ obtained by declaring $(x, x') \in R$ when $\mathsf{q}(x) = \mathsf{q}(x')$. Definability of θ and Q is then used to prove that R has finitely many equivalence classes. In this sense, Theorem 7.13 follows very closely the spirit of Theorem 7.8 for timed automata and provides another example of a quotient based abstraction.

When initial states $L \subset \mathbb{R}^n$ are to be described in the finite-state model S, we can refine the equivalence relation Q to an equivalence relation Q' respecting L, and apply Theorem 7.13 to Σ and Q'. The set of initial states on the finite-state system S can then be obtained as the set of all states related to the states of $S_{Q'}(\mathbb{R}^n, \theta)$ belonging to L.

Theorem 7.13 is a powerful result that can be specialized for different kinds of differential equations.

Corollary 7.14. *Let $\Sigma = (\mathbb{R}^n, f)$ be a dynamical system in which $f(x) = Ax$ is linear map represented by the matrix $A \in \mathbb{R}^{n\times n}$. If the eigenvalues of A are purely real, or if the matrix A is diagonalizable and its eigenvalues are purely imaginary, then for any definable finite equivalence relation Q on $\mathbb{R}^n$ there exists a finite-state system bisimilar to $S_Q(\Sigma)$.*

Proof. It is known from classical results [AM97] that the solution of a linear differential equation $d\xi/dt = A\xi$ with initial condition $x \in \mathbb{R}^n$ is given by $\xi(t) = e^{At}x = \sum_{i=0}^{\infty} \frac{1}{i!} A^i t^i x$. When the eigenvalues of A are real, for every $i = 1, 2, \ldots, n$, $\xi_i(t)$ is a semi-exponential-algebraic function of t and $x_1, x_2, \ldots, x_n$, see [AM97] for a detailed description of $\xi_i(t)$. Therefore, θ is definable and the result follows from Theorem 7.13. When the matrix A is diagonalizable and its eigenvalues are purely imaginary, then $\xi_i(t)$ can be written in terms of the analytic function sin. It follows from periodicity of sin that we can restrict its argument to $[-\pi, \pi]$ without altering its image. The restriction of sin induces a restriction θ' of θ which is definable in the order minimal structure containing restricted analytic functions. Since this restriction leads to a system $S_Q(\mathbb{R}^n, \theta')$ that is bisimilar to $S_Q(\mathbb{R}^n, \theta)$, the result follows by a straightforward adaptation of Theorem 7.13 applicable to $S_Q(\mathbb{R}^n, \theta')$. □

Example 7.15. A typical problem in control engineering is to design feedback control laws forcing the solutions of a particular differential equation, modeling the physical process being controlled, to converge to some desired operating point or region. We consider the following linear dynamical system Σ on $\mathbb{R}^2$:

$$f(x) = \begin{bmatrix} -7 & 3 \\ -3 & 0 \end{bmatrix} \begin{bmatrix} x_1 \\ x_2 \end{bmatrix} \tag{7.8}$$

which describes the result of applying a stabilizing feedback controller to a linear system. Although the solutions of the differential equation defined by f are guaranteed to converge to the origin, we would like to verify that trajectories do not violate the operating envelope by entering the set of unsafe states defined by:

$$B = \{(x_1, x_2) \in \mathbb{R}^2 \mid x_2 \leq -4 \ \vee \ x_2 \geq 4)\}$$

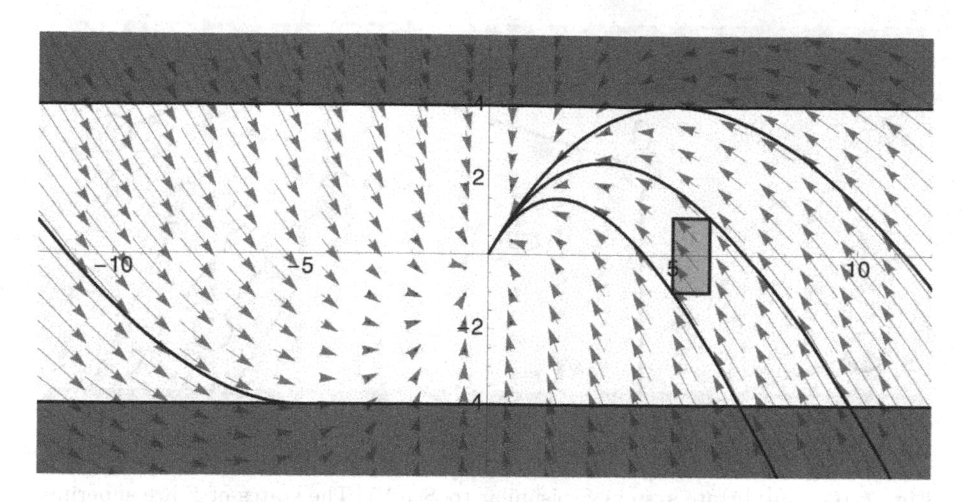

Fig. 7.9. Finite partition defined by an equivalence relation respecting continuous flows, the set of initial states L, and the set of unsafe states B. The set L is light-colored while the set B is dark-colored. The continuous-time dynamics defined by smooth map (7.8) is superimposed in gray.

when the initial conditions belong to the set:

$$L = \{(x_1, x_2) \in \mathbb{R}^2 \mid 5 \leq x_1 \leq 6 \ \wedge \ -1 \leq x_2 \leq 1\}.$$

The verification problem is specified by Σ and by the equivalence relation Q on $\mathbb{R}^2$ defined by the equivalence classes B, L, and $\mathbb{R}^2 \backslash (B \cup L)$. Noting that the eigenvalues of the matrix A defining the dynamics of Σ are purely real, we can apply Corollary 7.14 to conclude the existence of a finite-state system S bisimilar to $S_Q(\Sigma)$. The partition of $\mathbb{R}^2$ defining the states of S is depicted in Figure 7.9. The resulting finite-state system S is shown in Figure 7.10 with the states of S superimposed onto the equivalence classes they represent. The state in the center of the figure corresponds to the equivalence class containing only the state $\{(0, 0)\}$. By analyzing the finite-state system S we see that it is not possible to reach the set B from L and thus $S_Q(\Sigma)$ is safe. $\triangleleft$

The conditions in Corollary 7.14 are tight in the sense that there exist systems that do not satisfy these conditions and for which no finite-state bisimilar system exists. One such example is given by the dynamical system on $\mathbb{R}^2$ defined by:

$$f(x) = \begin{bmatrix} 0.1 & -1 \\ 1 & 0.1 \end{bmatrix} \begin{bmatrix} x_1 \\ x_2 \end{bmatrix}. \tag{7.9}$$

The corresponding A matrix has eigenvalues $0.1 \pm i$ and the solution to the differential equation defined by (7.9) is:

$$\theta(x, t) = e^{0.1t} \begin{bmatrix} \cos(t) & -\sin(t) \\ \sin(t) & \cos(t) \end{bmatrix} \begin{bmatrix} x_1 \\ x_2 \end{bmatrix}.$$

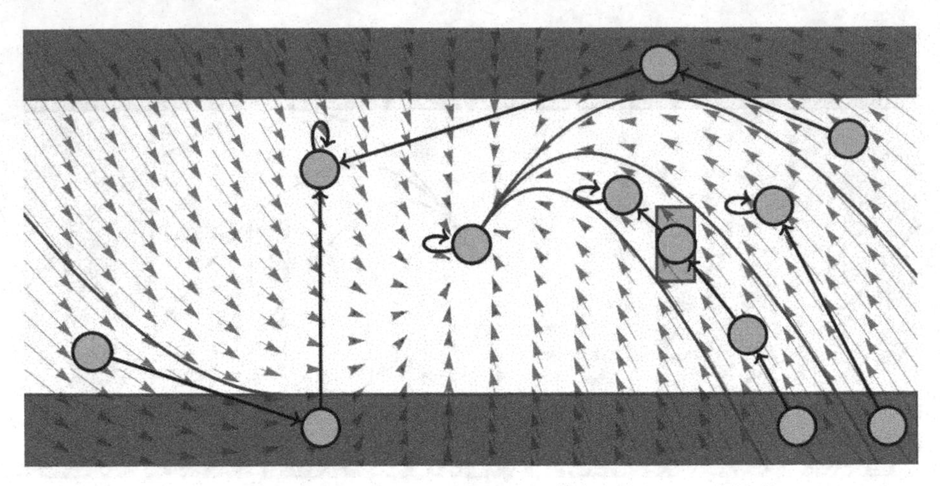

Fig. 7.10. Finite-state system S bisimilar to $S_Q(\Sigma)$. The states of S are superimposed on the equivalence classes they correspond to. The state in the center of the figure corresponds to the equivalence class consisting of the single state $(0, 0)$.

Solutions are spirals that unwind away from the origin as shown in Figure 7.11. Consider the partition of $\mathbb{R}^2$ defined by the sets:

$$P_1 = \{(x_1, 0) \in \mathbb{R}^2 \mid 0 < x_1 < 1\}$$
$$P_2 = \{(x_1, 0) \in \mathbb{R}^2 \mid -1 < x_1 < 0\}$$
$$P_3 = \mathbb{R}^2 \backslash P_1 \cup P_2$$

and let Q be the corresponding equivalence relation. Any system that is bisimilar to $S_Q(\Sigma)$ needs to distinguish between the points of P_2 that visit P_1 a

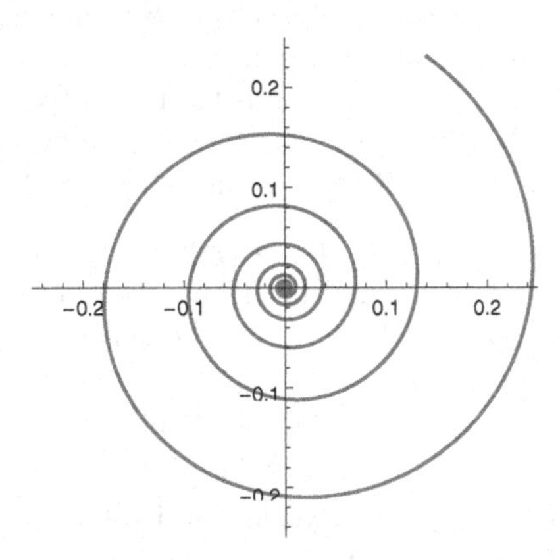

Fig. 7.11. Solution of the differential equation defined by (7.9) with initial condition $x_1 = 0$, and $x_2 = 0.001$.

different number $k \in \mathbb{N}_0$ of times. If we denote by P_{2k} the set of points of P_2 that visit P_1 k times, a simple computation shows that:

$$P_{2k} = \left\{(x_1, 0) \in \mathbb{R}^2 \mid e^{-\frac{(2k-1)\pi}{10}} < x_1 < 0\right\}.$$

Since these sets are all different for different values of k, we conclude that any system bisimilar to $S_Q(\Sigma)$ would necessarily have an infinite number of states and thus cannot be finite-state.

Differential equations in strict feed-forward form with a definable f also have definable solutions.

Corollary 7.16. *Let $\Sigma = (\mathbb{R}^n, f)$ be a dynamical system in which f is in strict feed-forward form, i.e., f is of the form:*

$$\begin{aligned} f_1(x) &= f_1(x_2, \ldots, x_n) \\ &\vdots \\ f_{n-2}(x) &= f_{n-2}(x_{n-1}, x_n) \\ f_{n-1}(x) &= f_{n-1}(x_n) \\ f_n(x) &= f_n \end{aligned}$$

with f_i definable for $i = 1, 2 \ldots, n$. Then, for any definable finite equivalence relation Q on $\mathbb{R}^n$ there exists a finite-state system bisimilar to $S_Q(\Sigma)$.

Proof. The differential equation $d\xi/dt = f(\xi)$ can be solved by successive integration and substitution. We start by integrating f_n to obtain:

$$\xi_n(t) = \xi_n(0) + \int_0^t f_n d\tau = \xi_n(0) + f_n t.$$

Then we substitute ξ_n into f_{n-1} to obtain $f_{n-1}(\xi_n(t), t)$ which can also be explicitly integrated, substituted in f_{n-2}, integrated, and so on. Since $\int_0^t f_n(\tau)d\tau$ is definable for a definable function f_n, see [Spe99], it follows that θ is definable. Therefore the result follows from Theorem 7.13. □

For general hybrid dynamical systems, refining a relation that already respects the continuous-time dynamics to a relation respecting reset maps, leads to a relation that no longer respects the continuous-time dynamics. Similarly, refining a relation that already respects the reset maps to a relation respecting the continuous-time dynamics, leads to a relation that no longer respects the reset maps. Therefore, there is no guarantee that a finite equivalence relation respecting both continuous-time dynamics and reset maps exists, even if all the data is definable. Nevertheless, Theorem 7.13 can still be used, in conjunction with Lemma 7.7, to show existence of finite-state bisimulations of hybrid dynamical systems with constant reset maps.

Corollary 7.17. *Let Σ be a hybrid dynamical system:*

$$\Sigma = (S_a, \{\mathrm{In}_{x_a}\}_{x_a \in X_a}, \{\mathrm{Gu}_{t_a}\}_{t_a \in \underset{a}{\longrightarrow}}, \{\mathrm{Re}_{t_a}\}_{t_a \in \underset{a}{\longrightarrow}}, \{f_{x_a}\}_{x_a \in X_a}).$$

satisfying the following properties:

1. Gu_{t_a} *is a definable set for every* $t_a \in \underset{a}{\longrightarrow}$ *;*
2. In_{x_a} *in a definable set for every* $x_a \in X_a$*;*
3. $(\mathrm{In}_{x_a}, f_{x_a})$ *is complete dynamical system with definable solutions for every* $x_a \in X_a$*;*
4. *the maps* Re_{t_a} *are constant for every* $t_a \in \underset{a}{\longrightarrow}$ *.*

For any collection of finite and definable equivalence relations $Q = \{Q_{x_a}\}_{x_a \in X_a}$ *with* $Q_{x_a} \subseteq \mathrm{In}_{x_a} \times \mathrm{In}_{x_a}$*, there exists a finite-state system bisimilar to* $S_Q(\Sigma)$*.*

Order minimality in an important tool in proving that certain classes of dynamical and hybrid dynamical systems admit finite-state bisimilar models. However, the computation of such abstractions remains a challenge since, in general, it requires the explicit knowledge of the solution ξ. In the following sections we describe different abstractions techniques that do not require the knowledge of ξ.

7.4 Sign based abstractions

We now describe a very different technique for the construction of finite-state abstractions based on a judicious analysis of the sign of real-valued functions on the state space. The underlying idea is quite simple and can be explained as follows. Consider a dynamical system $\Sigma = (\mathbb{R}^n, f)$ and let $p : \mathbb{R}^n \to \mathbb{R}$ be a smooth real-valued function. Function p induces a ternary partition of $\mathbb{R}^n$ defined by:

$$\begin{aligned} p_+ &= \{x \in \mathbb{R}^n \mid p(x) > 0\}, \\ p_0 &= \{x \in \mathbb{R}^n \mid p(x) = 0\}, \\ p_- &= \{x \in \mathbb{R}^n \mid p(x) < 0\}. \end{aligned}$$

Composing p with ξ we obtain $p \circ \xi : \mathbb{R} \to \mathbb{R}$, a smooth function of time. Continuity of $p \circ \xi$ tells us that if $p \circ \xi$ is negative for some $\tau_1 \in \mathbb{R}$ and positive for some $\tau_2 > \tau_1$, there must exist a $\tau \in]\tau_1, \tau_2[$ such that $p \circ \xi$ is zero at τ. We thus see that continuity imposes some restrictions on the different sequences of signs that a smooth real-valued function of the state can assume. Such restrictions can be encoded in a finite-state symbolic model for Σ describing how the trajectories of Σ interact with the sets p_+, p_0 and p_-. These simple ideas can be developed into a very powerful abstraction technique that we now describe in more detail.

	g_1	g_2	g_3	g_4	g_5	g_6	g_7	g_8	g_9
p_1	1	1	1	0	−1	−1	−1	0	0
p_2	1	0	−1	−1	−1	0	1	1	0

Table 7.1. Sign conditions for the functions p_1 and p_2 in (7.10).

Given a finite collection $\mathcal{P} = \{p_i\}_{i \in I}$ of smooth real-valued functions on $\mathbb{R}^n$, $\{1, 0, -1\}^{\mathcal{P}}$ denotes the set of all functions $g : \mathcal{P} \to \{1, 0, -1\}$. We call the functions $g \in \{1, 0, -1\}^{\mathcal{P}}$ sign conditions since $g(p_i) \in \{1, 0, -1\}$ is regarded as a sign for $p_i \in \mathcal{P}$. Each sign condition g defines a subset of $\mathbb{R}^n$ denoted by $\langle g \rangle$ and defined by:

$$\langle g \rangle = \bigcap_{i \in I} \{x \in \mathbb{R}^n \mid \text{sign}(p_i(x)) = g(p_i)\}.$$

A point $x \in \mathbb{R}^n$ belongs to $\langle g \rangle$ if for every function $p_i : \mathbb{R}^n \to \mathbb{R}$, the sign of $p_i(x)$ agrees with $g(p_i)$ in the sense that $\text{sign}(p_i(x)) = g(p_i)$. The collection of sets $\langle g \rangle$ with $g \in \{1, 0, -1\}^{\mathcal{P}}$ defines a partition of $\mathbb{R}^n$ and we denote the corresponding equivalence relation by P. We extend the notation $\langle g \rangle$ to sets of functions $G = \{g_1, g_2, \ldots, g_k\}$ by $\langle G \rangle = \bigcup_{i=1,\ldots,k} \langle g_i \rangle$.

Example 7.18. Let $n = 2$ and consider the functions:

$$p_1(x) = x_1, \qquad p_2(x) = x_2. \tag{7.10}$$

There are $3^2 = 9$ sign conditions as shown in Table 7.1. The sign condition g_3 defines the set $\langle g_3 \rangle$ consisting of all the points $(x_1, x_2) \in \mathbb{R}^2$ such that $\text{sign}(p_1) = \text{sign}(x_1) = 1$ and $\text{sign}(p_2) = \text{sign}(x_2) = -1$. This corresponds to the points satisfying the constraints $x_1 > 0$ and $x_2 < 0$. $\triangleleft$

The key idea underlying the construction of sign based abstractions is the interplay between the Lie derivative of p_i along f and the sign of p_i. The Lie derivative of a smooth function $p : \mathbb{R}^n \to \mathbb{R}$ along f is the function:

$$L_f p = \sum_{i=1}^{n} \frac{\partial p}{\partial x_i} f_i$$

which can also be expressed as:

$$L_f p\,(x) = \left.\frac{d}{dt}\right|_{t=0} p \circ \xi_x$$

with ξ_x the solution of $(\mathbb{R}^n, f)$. We also use $L_f^k p$ with $k \in \mathbb{N}$ to denote the kth Lie derivative of p along f defined by:

$$L_f^{k+1} p = L_f(L_f^k p), \qquad L_f^0 p = p.$$

If the inequality $L_f p_i(x) > 0$ is satisfied at a point $x = \xi_x(0)$ for which $p_i(x) = 0$, positivity of this derivative implies $p_i \circ \xi_x(\varepsilon) > 0$ for sufficiently small $\varepsilon > 0$. This can be restated as: if $\xi_x(0)$ belongs to the set $(\mathrm{sign} \circ p_i)^{-1}(0)$, then $\xi_x(\varepsilon)$ belongs to the set $(\mathrm{sign} \circ p_i)^{-1}(1)$. We can thus use the valuable information provided by the Lie derivatives of the functions p_i along f to build a model constraining the evolution of the signs of the functions p_i.

Definition 7.19. *Let $\Sigma = (\mathbb{R}^n, f)$ be a dynamical system, let $\mathcal{P} = \{p_i\}_{i \in I}$ be a finite collection of smooth real-valued functions on $\mathbb{R}^n$, and let P be the equivalence relation defined by $\mathcal{P}$. For any set of initial states $L \subseteq \mathbb{R}^n$, the finite-state system induced by $S(\Sigma)$, $\mathcal{P}$, and L, denoted by $S_{\mathcal{P}L}(\Sigma)$, consists of:*

- $X = \{1, 0, -1\}^{\mathcal{P}}$;
- $X_0 = \{g \in X \mid \langle g \rangle \cap L \neq \varnothing\}$;
- $U = \{*\}$;
- $g \xrightarrow{*} g'$ *if for every $i \in I$ any of the following holds:*
  1. *$g(p_i) = 1$ implies any of the following:*
     a) $\mathrm{sign}(L_f p_i(\langle g \rangle)) \subseteq \{1, 0\}$ *and* $g'(p_i) = 1$;
     b) $\mathrm{sign}(L_f p_i(\langle g \rangle)) \supseteq \{-1\}$ *and* $g'(p_i) \in \{1, 0\}$;
  2. *$g(p_i) = 0$ implies any of the following:*
     a) $\mathrm{sign}(L_f p_i(\langle g \rangle)) = \{1\}$ *and* $g'(p_i) = 1$;
     b) $\mathrm{sign}(L_f p_i(\langle g \rangle)) = \{-1\}$ *and* $g'(p_i) = -1$;
     c) $\mathrm{sign}(L_f p_i(\langle g \rangle)) = \{1, -1\}$ *and* $g'(p_i) \in \{1, -1\}$;
     d) $\mathrm{sign}(L_f p_i(\langle g \rangle)) \supseteq \{0\}$ *and* $g'(p_i) \in \{1, 0, -1\}$;
  3. *$g(p_i) = -1$ implies any of the following:*
     a) $\mathrm{sign}(L_f p_i(\langle g \rangle)) \supseteq \{1\}$ *and* $g'(p_i) \in \{0, -1\}$;
     b) $\mathrm{sign}(L_f p_i(\langle g \rangle)) \subseteq \{0, -1\}$ *and* $g'(p_i) = -1$;
- $Y = \mathbb{R}^n / P$;
- $H(g) = \langle g \rangle$.

Before proceeding further, we illustrate the construction of $S_{\mathcal{P}L}(\Sigma)$.

Example 7.20. Consider the linear dynamical system on $\mathbb{R}^2$ defined by:

$$f(x) = \begin{bmatrix} x_2 \\ -x_1 \end{bmatrix}. \tag{7.11}$$

The trajectories of this system revolve around the origin under a circular clock-wise motion as depicted in Figure 7.12. We take $\mathcal{P} = \{p_1, p_2\}$, with p_1 and p_2 as defined in (7.10), and $L = \{x \in \mathbb{R}^2 \mid x_1 > 0 \wedge x_2 > 0\}$. The state set X is defined by the $3^2 = 9$ functions represented in Table 7.1. The transition relation of $S_{\mathcal{P}L}(\Sigma)$ is constructed by adding transitions $g \xrightarrow{*} g'$ according to the rules in Definition 7.19. We illustrate the construction for the transitions originating from g_1 and g_2. We first compute:

$$L_f p_1 = \frac{\partial p_1}{\partial x_1} f_1 + \frac{\partial p_1}{\partial x_2} f_2 = x_2, \qquad L_f p_2 = \frac{\partial p_2}{\partial x_1} f_1 + \frac{\partial p_2}{\partial x_2} f_2 = -x_1.$$

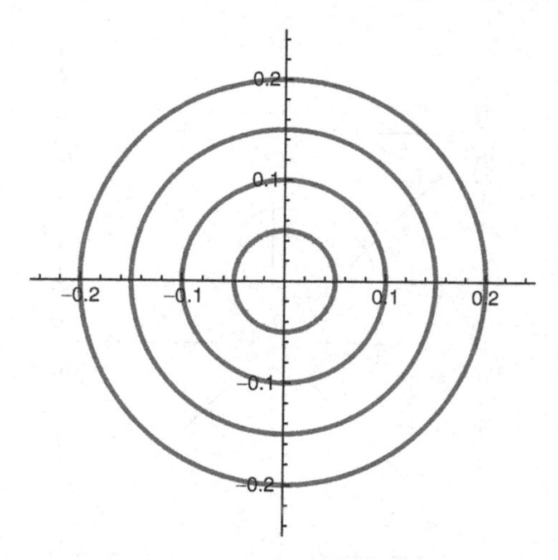

Fig. 7.12. Trajectories of the dynamical system defined by (7.11) with initial conditions $(x_1, x_2) = (0, 0.05)$, $(x_1, x_2) = (0, 0.1)$, $(x_1, x_2) = (0, 0.15)$, and $(x_1, x_2) = (0, 0.2)$.

On the set $\langle g_1 \rangle = \{(x_1, x_2) \in \mathbb{R}^2 \mid x_1 > 0 \wedge x_2 > 0\}$ we have $L_f p_1 > 0$ and $L_f p_2 < 0$. Since p_1 is positive in $\langle g_1 \rangle$ and $L_f p_1 > 0$, the successor g' of g_1 satisfies $g'(p_1) = 1$. Moreover, as $p_2 > 0$ in $\langle g_2 \rangle$ and $L_f p_2 < 0$, g' also satisfies $g'(p_2) \in \{1, 0\}$. The only elements $g' \in X$ satisfying the constraints $g'(p_1) = 1$ and $g'(p_2) \in \{1, 0\}$ are g_1 and g_2. Therefore, there are only the following two transitions originating from g_1:

$$g_1 \xrightarrow{*} g_1, \qquad g_1 \xrightarrow{*} g_2.$$

Similarly, on the set $\langle g_2 \rangle$ we have $L_f p_1 = 0$ and $L_f p_2 < 0$ implying that the successor g' of g_2 satisfies $g'(p_1) = 1$ and $g'(p_2) = -1$. Consequently, the only transition originating from g_2 is:

$$g_2 \xrightarrow{*} g_3.$$

The remaining transitions can be similarly constructed resulting in the system $S_{\mathcal{P}L}(\Sigma)$ depicted in Figure 7.13. Although $S_{\mathcal{P}L}(\Sigma)$ simulates $S_{\mathcal{P}L}(\Sigma)$, $S_{\mathcal{P}L}(\Sigma)$ is not bisimilar to $S_{\mathcal{P}L}(\Sigma)$ since, *e.g.*, the transitions $g_i \xrightarrow{*} g_i$ with $i \in \{1, 3, 5, 7\}$ do not correspond to transitions in $S_{\mathcal{P}L}(\Sigma)$. $\triangleleft$

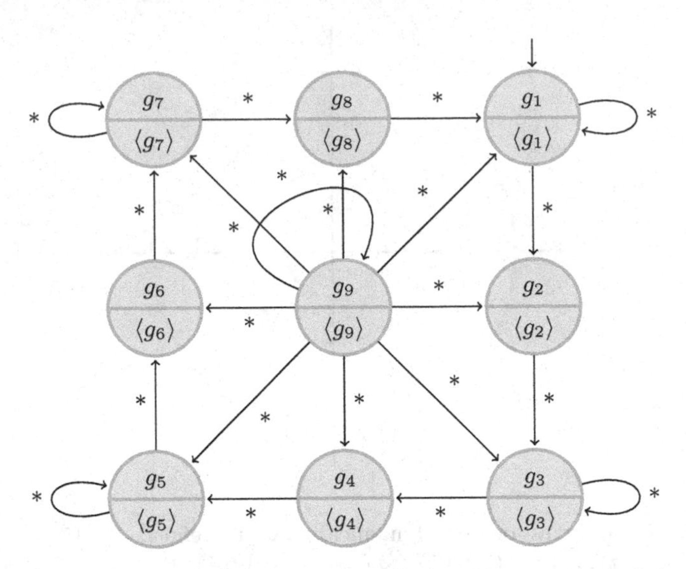

Fig. 7.13. Finite-state sign based system $S_{\mathcal{P}L}(\Sigma)$ simulating the infinite-state system $S_{PL}(\Sigma)$.

The system $S_{\mathcal{P}L}(\Sigma)$ is a finite-state abstraction of $\S_{PL}(\Sigma)$ in the following sense.

Proposition 7.21. *Let* $\Sigma = (\mathbb{R}^n, f)$ *be a dynamical system, and let* $\mathcal{P}$ *be a finite collection of smooth real-valued functions on* $\mathbb{R}^n$. *For any set of initial states* $L \subseteq \mathbb{R}^n$, *the relation* $R \subseteq \mathbb{R}^n \times \{1, 0, -1\}^{\mathcal{P}}$ *defined by:*

$$(x, g) \in R \quad \text{if} \quad x \in \langle g \rangle$$

is a simulation relation from $S_{PL}(\Sigma)$ *to* $S_{\mathcal{P}L}(\Sigma)$.

Proof. By construction, the first and second conditions in the definition of simulation relation are satisfied. Let now $(x, g) \in R$, assume that $x \xrightarrow{\tau} x'$ in $S_{PL}(\Sigma)$ and recall that this implies the existence of a trajectory ξ satisfying $\xi(0) = x$ and $\xi(\tau) = x'$. Consider any $p_i \in \mathcal{P}$. We only provide the details for the case where $g(p_i) = 1$ since all the other cases are similar. By definition of the equivalence relation P defined by $\mathcal{P}$, and by definition of $S_{PL}(\Sigma)$, three situations can occur: $\text{sign}(p_i \circ \xi(t)) = 1$ for $t \in [0, \tau]$; $\text{sign}(p_i \circ \xi(t)) = 1$ for $t \in [0, \varepsilon[$ and $\text{sign}(p_i \circ \xi(t)) = 0$ for $t \in [\varepsilon, \tau]$; or $\text{sign}(p_i \circ \xi(t)) = 1$ for $t \in [0, \varepsilon]$ and $\text{sign}(p_i \circ \xi(t)) = 0$ for $t \in]\varepsilon, \tau]$.

First situation: according to Definition 7.19, independently of the sign of $L_f p_i(\langle g \rangle)$, there exists a transition $g \xrightarrow{*} g'$ in $S_{\mathcal{P}L}(\Sigma)$ with $g'(p_i) = 1$ and thus $(x', g') \in R$.

Second and third situations: in both cases cases, smoothness of $p_i \circ \xi$ implies the existence of $t' \in [0, \varepsilon[$ such that:

$$\left.\frac{d}{dt}\right|_{t=t'} p_i \circ \xi < 0.$$

Therefore, $L_f p_i(x'') < 0$ for $x'' = \xi(t')$ and thus $\{-1\} \subseteq \mathrm{sign}(L_f p_i(\langle g \rangle))$. It then follows from Definition 7.19 the existence of a transition $g \xrightarrow{*} g'$ in $S_{\mathcal{P}L}(\Sigma)$ with $g'(p_i) = 0$, hence $(x', g') \in R$.

Using the same argument for the remaining cases we conclude that R is a simulation relation as desired. □

The relationship $S_{PL}(\Sigma) \preceq_S S_{\mathcal{P}L}(\Sigma)$ is quite useful in practice since the construction of $\Sigma_{\mathcal{P}L}(\Sigma)$ only requires the knowledge of the signs of Lie derivatives. By contrast, the construction of $S_{PL}(\Sigma)$ requires knowledge of the trajectories of Σ.

Example 7.22. Sign based abstractions can also be used to verify the system in Example 7.15. Consider the set $\mathcal{P}$ consisting of the single function:

$$p(x) = 7x_1^2 - 6x_1x_2 + 28x_2^2 - 320.$$

The reason for choosing this function will become apparent in the next section where we discuss barrier certificates. For now, we compute $S_{\mathcal{P}L}(\Sigma)$. We have three states $g_1(p) = 1$, $g_2(p) = 0$, and $g_3(p) = -1$. The set of initial states is $X_0 = \{g_3\}$ since $L \subset \langle g_3 \rangle$. To construct the transition relation we note that $L_f p(x) < 0$ for any $x \in \mathbb{R}^2$. According to Definition 7.19 the transition relation consists of the transitions:

$$g_1 \xrightarrow{*} g_1, \quad g_1 \xrightarrow{*} g_2, \quad g_2 \xrightarrow{*} g_3, \quad g_3 \xrightarrow{*} g_3.$$

The resulting abstraction is depicted in Figure 7.14, clearly showing that g_1 is not reachable in $S_{\mathcal{P}L}(\Sigma)$. Since $S_{\mathcal{P}L}(\Sigma)$ simulates $S_{PL}(\Sigma)$ and $B \subset \langle g_1 \rangle$, the system is safe. ◁

The rules used to construct the transition relation in $S_{\mathcal{P}L}(\Sigma)$ are conservative whenever the sign of $L_f p_i$ is not constant on a given set $\langle g \rangle$ since in this case $g'(p_i)$ is not uniquely determined. A less conservative abstraction can be obtained by adding $L_f p_i$ as a new function to the set $\mathcal{P}$ if the sign of $L_f p_i$

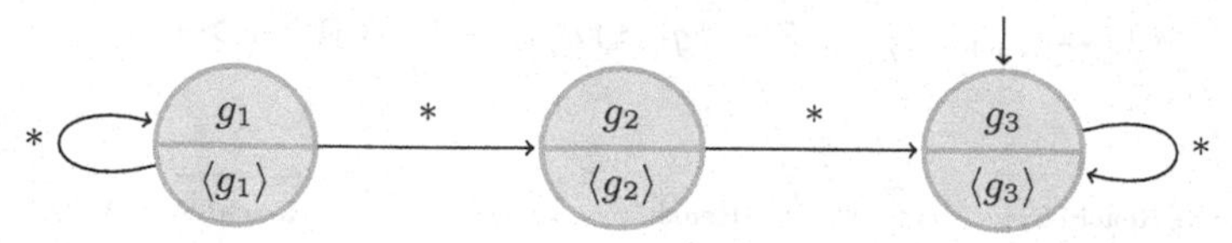

Fig. 7.14. Sign based abstraction $S_{\mathcal{P}L}(\Sigma)$ of $S_{PL}(\Sigma)$.

is not constant on $\langle g \rangle$. Termination of this closure or saturation process, consisting of adding Lie derivatives to $\mathcal{P}$, leads to a set of functions $\mathcal{P}$ for which $S_{\mathcal{P}L}(\Sigma)$ is a tight abstraction of $S_{PL}(\Sigma)$ as measured by the inclusions:

$$\cup_{Z \in \text{Reach}(S_{PL}(\Sigma))} Z \subseteq \cup_{Z \in \text{Reach}(S_{\mathcal{P}L}(\Sigma))} Z \subseteq \overline{\cup_{Z \in \text{Reach}(S_{PL}(\Sigma))} Z}. \tag{7.12}$$

Before discussing further the preceding inclusions, we need to give meaning to the expression "termination of the saturation process". We say that a set $\mathcal{P}$ of smooth real-valued functions is closed with respect to the sign of the Lie derivative when for every $p \in \mathcal{P}$ and for every $k \in \mathbb{N}$, the sign of $L_f^k p$ is constant on every set $\langle g \rangle$ and can be determined from the knowledge of $g \in \{1, 0, -1\}^{\mathcal{P}}$.

Example 7.23. The set $\mathcal{P}$ used in Example 7.20 is closed with respect to the sign of the Lie derivative since $L_f p_1 = p_2$ and $L_f p_2 = -p_1$ imply $\text{sign}(L_f p_1) = \text{sign}(p_2)$ and $\text{sign}(L_f p_2) = -\text{sign}(p_1)$. Hence, by recursively using these equalities we can determine the sign of $L_f^k p_1$ and $L_f^k p_2$ for any $k \in \mathbb{N}$ as a function of the sign of p_1 and p_2. $\lhd$

Whenever the set of functions $\mathcal{P}$ is closed with respect to the the sign of the Lie derivative, the rules for defining the transition relation of $S_{\mathcal{P}L}(\Sigma)$ simplify since there is no ambiguity in the sign of $(L_f p_i)(\langle g \rangle)$. A careful analysis of these rules shows that the only transitions in $S_{\mathcal{P}L}(\Sigma)$ that might not correspond to transitions in $S_{PL}(\Sigma)$ are generalizations of the next example.

Example 7.24. Consider the dynamical system $\Sigma = (\mathbb{R}, -x)$ and the set $\mathcal{P}$ consisting of the just one function $p(x) = x$. It is easy to see that in this case the set $\mathcal{P}$ is closed with respect to the sign of the Lie derivative. Moreover, $\{1, 0, -1\}^{\mathcal{P}} = \{g_1, g_2, g_3\}$ with $g_1(p) = 1$, $g_2(p) = 0$ and $g_3(p) = -1$. On the set defined by $\langle g_1 \rangle$ we have $L_f p = -x < 0$. According to the rules in Definition 7.19 we have the following two transitions, in $S_{\mathcal{P}L}(\Sigma)$, from g_1:

$$g_1 \xrightarrow{*} g_2, \qquad g_1 \xrightarrow{*} g_1.$$

However, for any initial condition $x > 0$ we know that the solution $\xi(t) = e^{-t}x$ of $\dot{\xi} = -\xi$ satisfying $\xi(0) = x$ will never reach the set $\langle g_2 \rangle$ and thus the transition $g_1 \xrightarrow{*} g_2$ does not correspond to a transition in $S_{PL}(\Sigma)$.

Nevertheless, when $L = \langle g_1 \rangle$ we have:

$$\cup_{Z \in \text{Reach}(S_{PL}(\Sigma))} Z = \langle g_1 \rangle = \{x \in \mathbb{R} \mid x > 0\}$$

and:

$$\cup_{Z \in \text{Reach}(S_{\mathcal{P}L}(\Sigma))} Z = \langle g_1 \rangle \cup \langle g_2 \rangle = \{x \in \mathbb{R} \mid x \geq 0\},$$

hence:

$$\cup_{Z \in \text{Reach}(S_{PL}(\Sigma))} Z \subseteq \cup_{Z \in \text{Reach}(S_{\mathcal{P}L}(\Sigma))} Z \subseteq \overline{\cup_{Z \in \text{Reach}(S_{PL}(\Sigma))} Z}.$$

$\lhd$

The containment (7.12) holds whenever $\mathcal{P}$ is closed with respect to the sign of Lie derivatives. This fact, formally stated below, can be proved by meticulously generalizing Example 7.24.

Theorem 7.25. *Consider a dynamical system $\Sigma = (\mathbb{R}^n, f)$, let $\mathcal{P}$ be a finite collection of smooth real-valued functions on $\mathbb{R}^n$, and let P be the equivalence relation defined by $\mathcal{P}$. If L is any set of initial states that is a union of equivalence classes of P, and if $\mathcal{P}$ is closed with respect to the sign of the Lie derivative, then:*

$$\cup_{Z \in \text{Reach}(S_{PL}(\Sigma))} Z \subseteq \cup_{Z \in \text{Reach}(S_{\mathcal{P}L}(\Sigma))} Z \subseteq \overline{\cup_{Z \in \text{Reach}(S_{PL}(\Sigma))} Z}. \tag{7.13}$$

The inclusions (7.13) show that $S_{\mathcal{P}L}(\Sigma)$ defines a tight abstraction of $S_{PL}(\Sigma)$. From Proposition 7.21 and Proposition 4.6 we know that:

$$\text{Reach}(S_{PL}(\Sigma)) \subseteq \text{Reach}(S_{\mathcal{P}L}(\Sigma))$$

from which follows directly:

$$\cup_{Z \in \text{Reach}(S_{PL}(\Sigma))} Z \subseteq \cup_{Z \in \text{Reach}(S_{\mathcal{P}L}(\Sigma))} Z$$

by noting that $\text{Reach}(S_{PL}(\Sigma))$ and $\text{Reach}(S_{\mathcal{P}L}(\Sigma))$ are sets of equivalence classes. Although the reverse containment:

$$\cup_{Z \in \text{Reach}(S_{PL}(\Sigma))} Z \supseteq \cup_{Z \in \text{Reach}(S_{\mathcal{P}L}(\Sigma))} Z$$

cannot be guaranteed, a state belonging to $\cup_{Z \in \text{Reach}(S_{\mathcal{P}L}(\Sigma))} Z$ that fails to belong to $\cup_{Z \in \text{Reach}(S_{PL}(\Sigma))} Z$ must belong to the boundary of $\cup_{Z \in \text{Reach}(S_{PL}(\Sigma))} Z$ which is a "thin" set. This can also be expressed by the equality:

$$\text{int} \cup_{Z \in \text{Reach}(S_{PL}(\Sigma))} Z = \text{int} \cup_{Z \in \text{Reach}(S_{\mathcal{P}L}(\Sigma)} Z.$$

Theorem 7.25 requires the collection of functions $\mathcal{P}$ to be closed with respect to the sign of Lie derivative. In concrete applications, one starts with a set of functions $\mathcal{P}$ defined by the verification problem to be solved. Since the set $\mathcal{P}$ may not be closed with respect to the sign of the Lie derivative, one can enlarge $\mathcal{P}$ by adding the functions $L_f p$. Unfortunately, this saturation or closure procedure is not guaranteed to terminate except in some special cases. When the dynamics is linear, *i.e.*, $f(x) = Ax$ for some matrix A, there are several obvious choices for the functions p. Some examples include $p(x) = v^T x$ with v an eigenvector of A^T. In this case:

$$L_f p = v^T A x = (A^T v)^T x = (\lambda v)^T x = \lambda p$$

and thus $\text{sign}(L_f^k p) = \text{sign}(\lambda^k)\text{sign}(p)$. If A is nilpotent, then for any affine function p there exists a $l \in \mathbb{N}$ such that $L_f^k p = 0$ for all $k \geq l$.

It is important to emphasize that even if the saturation process does not terminate, $S_{\mathcal{P}L}(\Sigma)$ is always guaranteed to simulate $S_{PL}(\Sigma)$ and can be detailed enough to verify the desired properties.

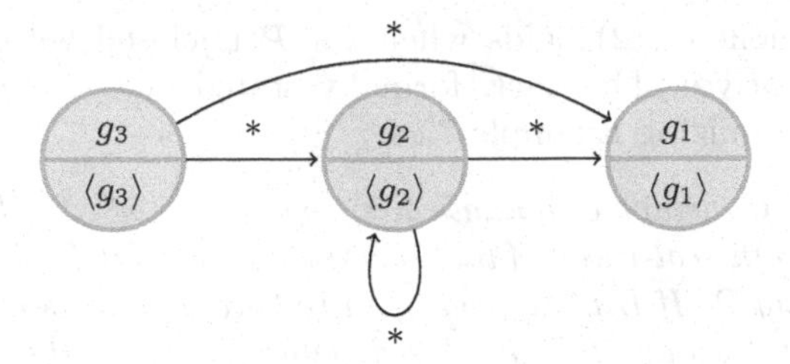

Fig. 7.15. Sign based abstraction of $S_P([0,\pi],-1)$ for Example 7.26.

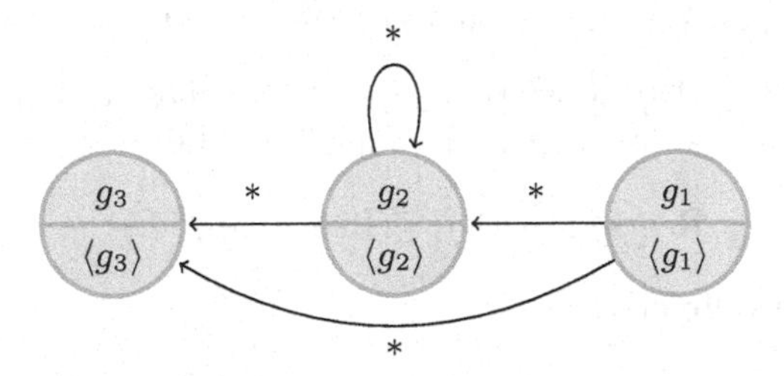

Fig. 7.16. Sign based abstraction of $S_P([0,\pi],1)$ for Example 7.26.

Sign based abstractions can also be used to verify hybrid dynamical systems. Given a hybrid dynamical system Σ, we can use Definition 7.19 to construct a finite-state system S_{x_a}, for every $x_a \in X_a$, simulating the continuous-time dynamics in each finite state of Σ. These abstractions can then be weaved together into an abstraction of Σ by adding discrete transitions between the systems S_{x_a} simulating the discrete transitions of Σ. This construction is illustrated in the next example.

Example 7.26. We revisit Example 7.4 and in particular the hybrid dynamical system Σ, depicted in Figure 7.1, modeling the windshield wiper. If $\mathcal{P}$ consists of the functions $p_1(x) = x$ and $p_2(x) = x - \frac{\pi}{2}$ there are only three nonempty sign conditions. The corresponding sets are:

$$\begin{aligned}\langle g_1\rangle &= \{x \in [0,\pi] \mid x = 0 \ \wedge \ x - \frac{\pi}{2} < 0\}\\ \langle g_2\rangle &= \{x \in [0,\pi] \mid x > 0 \ \wedge \ x - \frac{\pi}{2} < 0\}\\ \langle g_3\rangle &= \{x \in [0,\pi] \mid x > 0 \ \wedge \ x - \frac{\pi}{2} = 0\}.\end{aligned}$$

Following Definition 7.19 we obtain the sign based abstraction $S_{\mathcal{P}}([0,\pi],-1)$ of $S_P([0,\pi],-1)$, represented in Figure 7.15, and the sign based abstraction $S_{\mathcal{P}}([0,\pi],1)$ of $S_P([0,\pi],1)$, represented in Figure 7.16. The finite-state systems $S_{\mathcal{P}}([0,\pi],-1)$ and $S_{\mathcal{P}}([0,\pi],1)$ can now be combined into a single finite-state system simulating $S_Q(\Sigma)$ with $Q = \{Q_{\mathtt{right}}, Q_{\mathtt{left}}\}$ where $Q_{\mathtt{left}} = P = Q_{\mathtt{right}}$. This is done by adding transitions simulating the discrete transitions of $S_Q(\Sigma)$ as shown in Figure 7.17. $\triangleleft$

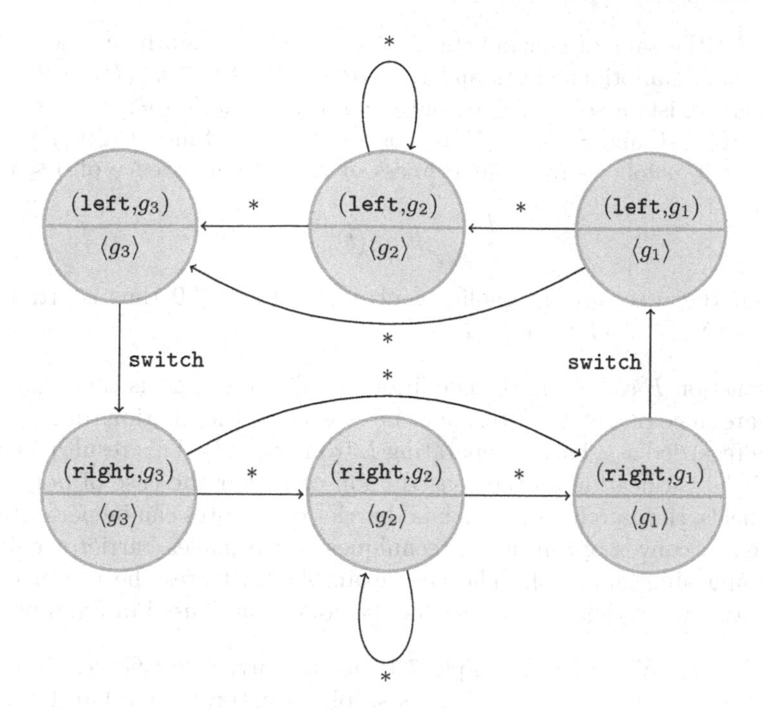

Fig. 7.17. Sign based abstraction of $S_P(\Sigma)$ for Example 7.26.

7.5 Barrier certificates

Constructing finite-state bisimulations of $S_{QL}(\Sigma)$, when Σ is a dynamical or hybrid dynamical system, is a difficult task even when they are known to exist. However, safety verification problems can often be solved without computing very detailed finite-state systems abstracting $S_{QL}(\Sigma)$. Recall that the safety verification problem asks if given a system S, which we take as $S = S_{QL}(\Sigma)$, and given a set of unsafe outputs $B \subseteq Y$, wether $\text{Reach}(S_{QL}(\Sigma)) \cap \pi_Q(B) = \varnothing$. The next result provides a very simple and elegant sufficient solution for an affirmative answer to the safety verification problem.

Theorem 7.27. *Let $\Sigma = (\mathbb{R}^n, f)$ be a dynamical system, let $L \subseteq \mathbb{R}^n$ be a set of initial states, let $B \subseteq \mathbb{R}^n = Y$ be a set of unsafe outputs, and let Q be a finite equivalence relation on $\mathbb{R}^n$ respecting L and $\pi_Q^{-1}(B)$. If there exists a smooth function $E : \mathbb{R}^n \to \mathbb{R}$ satisfying:*

1. *$E(x) \leq 0$ for $x \in L$;*
2. *$E(x) > 0$ for $x \in \pi_Q^{-1}(B)$;*
3. *$(L_f E)(x) \leq 0$ for $x \in \mathbb{R}^n$;*

then $\text{Reach}(S_{QL}(\Sigma)) \cap \pi_Q(B) = \varnothing$.

Proof. For the sake of contradiction assume that a smooth function E satisfying the assumptions exists and that $\mathrm{Reach}(S_L(\Sigma)) \cap \pi_Q(B) \neq \varnothing$. Then, there must exists a solution ξ of the differential equation $d\xi/dt = f(\xi)$ satisfying $\xi(0) \in L$ and $\xi(\tau) \in \pi_Q^{-1}(B)$ for some $\tau \in \mathbb{R}$. Since $E(\xi(0)) \leq 0$ and $E(\xi(\tau)) > 0$, it follows from smoothness of $E \circ \xi$ the existence of $0 \leq \tau' \leq \tau$ such that:

$$\left.\frac{d}{dt}\right|_{t=\tau'} E \circ \xi(t) > 0.$$

However, the preceding inequality contradicts $L_f E \leq 0$ thus contradicting $\mathrm{Reach}(S_L(\Sigma)) \cap \pi_Q(B) \neq \varnothing$. □

A function E satisfying the conditions in Theorem 7.27 is called a barrier certificate since the set $E^{-1}(0)$ cannot be crossed by the solutions of Σ and can thus be regarded as a barrier separating L from $\pi_Q^{-1}(B)$. This result places the onus of the verification on the construction of E. For the case of polynomial vector fields, the search for polynomial barrier certificates can be performed by resorting to convex optimization techniques which makes barrier certificates a very appealing approach. The next example illustrates the use of barrier certificates and explains the choice for the collection $\mathcal{P}$ used in Example 7.22.

Example 7.28. We revisit Example 7.15 using barrier certificates. Since the dynamical system in Example 7.15 is stable, a natural candidate for E is a Lyapunov like function. The following choice:

$$E(x) = 7x_1^2 - 6x_1x_2 + 28x_2^2 - 320 \tag{7.14}$$

satisfies all the conditions in Theorem 7.27 and thus proves that system Σ in Example 7.15 is safe. The zero level set of E is the ellipsoid represented in Figure 7.18 where it can be seen that it does separate L from $\pi_Q^{-1}(B)$. ◁

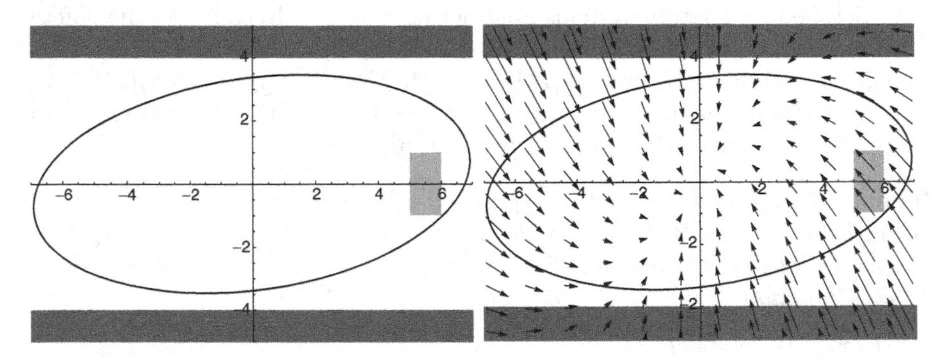

Fig. 7.18. Representation of the zero level set for the barrier certificate (7.14). The initial set is light-colored while the unsafe set is dark-colored. The continuous-time dynamics defined by the smooth map (7.8) is shown in the figure on the right.

7.6 Computation of reachable sets

A direct approach to safety verification problems is the computation of an over-approximation to the reachable set. When W is such an over-approximation, *i.e.*, $\text{Reach}(S_{QL}(\Sigma)) \subseteq W$, showing that $W \cap B = \varnothing$ is sufficient to conclude that $\text{Reach}(S_{QL}(\Sigma)) \cap B = \varnothing$. The computation of reachable sets is not only useful for verification purposes but also for the construction of some symbolic models for control discussed in Chapter 11.

Several different competing techniques for the computation of reachable sets are available in the literature. Here, we focus on a very simple but efficient technique applicable to linear dynamical systems. The starting point is a linear dynamical system $\Sigma = (\mathbb{R}^n, f)$ with $f(x) = Ax$, for some matrix $A \in \mathbb{R}^{n \times n}$, and a set of initial states $L \subseteq \mathbb{R}^n$. We use *zonotopes* to represent all the sets appearing in the computations. A zonotope is a convex polytope that is the image of the unit cube under an affine transformation. An equivalent definition will be more convenient for our purposes.

Definition 7.29 (Zonotope). *A* zonotope *Z is a set $Z \subseteq \mathbb{R}^n$ described by:*

$$Z = \left\{ x \in \mathbb{R}^n \mid x = c + \sum_{i=1}^{k} \lambda_i v_i, \quad -1 \leq \lambda_i \leq 1 \right\}$$

where $c, v_1, v_2, \ldots, v_k$ are vectors in $\mathbb{R}^n$.

We denote a zonotope Z by $Z = (c, < v_1, \ldots, v_k >)$ and note that the point c is the center of the zonotope. The class of zonotopes is closed under linear transformations and Minkowski sum.

Proposition 7.30. *Consider two zonotopes $Z_a = (c_a, < v_{a1}, \ldots, v_{ak} >)$ and $Z_b = (c_b, < v_{b1}, \ldots, v_{bl} >)$ in $\mathbb{R}^n$ and let $g(x) = Gx$ be a linear transformation. The following holds:*

- $Z_a + Z_b = (c_a + c_b, < v_{a1}, \ldots, v_{ak}, v_{b1}, \ldots, v_{bl} >)$;
- $g(Z_a) = (Gc_a, < Gv_{a1}, \ldots, Gv_{ak} >)$.

We now denote by $\mathcal{R}_\tau(Z)$ the set of points reached at time τ by trajectories of Σ starting in Z:

$$\mathcal{R}_\tau(Z) = \{x \in \mathbb{R}^n \mid x = \xi(\tau) \ \wedge \ \xi(0) \in Z\}.$$

and by $\mathcal{R}_{[\tau_1,\tau_2]}(Z)$ the set of points reached at time $\tau \in [\tau_1, \tau_2]$ by trajectories of Σ starting in Z:

$$\mathcal{R}_{[\tau_1,\tau_2]}(Z) = \bigcup_{\tau \in [\tau_1,\tau_2]} \mathcal{R}_\tau(Z).$$

The computation of $\mathcal{R}_\tau(Z)$ is straightforward when Z is a zonotope $Z = (c, < v_1, \ldots, v_k >)$. Since the solution of $d\xi/dt = A\xi$ with initial condition $x \in \mathbb{R}^n$ is $\xi(t) = e^{At}x$ and since $e^{A\tau}$ is a linear transformation we have:

$$\mathcal{R}_\tau(Z) = e^{A\tau}(Z) = (e^{A\tau}c, < e^{A\tau}v_1, \ldots, e^{A\tau}v_k >).$$

The computation of $\mathcal{R}_{[0,\tau]}(Z)$ involves a sequence of steps. First, we construct a zonotope Z_a containing Z and $e^{A\tau}(Z)$ by:

$$Z_a = \left(\frac{c + e^{A\tau}c}{2}, \left\langle \frac{v_1 + e^{A\tau}v_1}{2}, \ldots, \frac{v_k + e^{A\tau}v_k}{2}, \right.\right.$$
$$\left.\left. \frac{c - e^{A\tau}c}{2}, \frac{v_1 - e^{A\tau}v_1}{2}, \ldots, \frac{v_k - e^{A\tau}v_k}{2} \right\rangle\right). \quad (7.15)$$

The particular form of Z_a was chosen for technical reasons that will become apparent in the proof of Proposition 7.31. Although Z_a contains Z and $e^{A\tau}(Z)$, there is no guarantee that it contains $\mathcal{R}_{[0,\tau]}(Z)$. The next step is to inflate Z_a by computing:

$$Z_b = Z_a + \mathcal{C}_{\alpha_\tau} \quad (7.16)$$

where $\mathcal{C}_{\alpha_\tau}$ is the hyper-cube centered at zero and of radius α_τ. Note that $\mathcal{C}_{\alpha_\tau}$ is a zonotope since it can be written as $(0, < \alpha_\tau b_1, \ldots, \alpha_\tau b_n >)$ with b_i being the vector containing a one in position i and zeros elsewhere. The parameter α_τ is chosen so that Z_b is large enough to contain $\mathcal{R}_{[0,\tau]}(Z)$.

Proposition 7.31. *Let $\Sigma = (\mathbb{R}^n, f)$ be a dynamical system with $f(x) = Ax$ for $A \in \mathbb{R}^{n\times n}$ and let $Z \subseteq \mathbb{R}^n$ be a zonotope. If the parameter α_τ is chosen according to:*

$$\alpha_\tau = \left(e^{\|A\|_\infty \tau} - 1 - \tau\|A\|_\infty\right) \sup_{x\in Z} \|x\|_\infty$$

then $\mathcal{R}_{[0,\tau]}(Z) \subseteq Z_b$ where Z_b is the zonotope computed according to (7.16) and (7.15).

Proof. Let $x \in Z$ and $t \in [0, \tau]$. We use:

$$\widehat{\xi}(t) = \left(1 - \frac{t}{\tau}\right)x + \frac{t}{\tau}e^{A\tau}x \quad (7.17)$$

as an estimate of $\xi(t) = e^{At}x$. Note that for $t = 0$ and $t = \tau$ we have $\widehat{\xi}(0) = x = \xi(0)$ and $\widehat{\xi}(\tau) = e^{A\tau}x = \xi(\tau)$. For other values of $t \in [0, \tau]$ the error in this estimate is bounded by:

$$\|\xi(t) - \widehat{\xi}(t)\|_\infty = \left\|e^{At}x - \left(1 - \frac{t}{\tau}\right)x - \frac{t}{\tau}e^{A\tau}x\right\|_\infty$$
$$= \left\|\sum_{k=2}^{\infty} \frac{t(t^{k-1} - \tau^{k-1})}{k!} A^k x\right\|_\infty$$

$$
\begin{aligned}
&\leq \left\| \sum_{k=2}^{\infty} \frac{\tau^k}{k!} A^k x \right\|_\infty \\
&\leq \sum_{k=2}^{\infty} \frac{\tau^k}{k!} \|A\|_\infty^k \|x\|_\infty \\
&= \left(e^{\|A\|_\infty \tau} - 1 - \|A\|_\infty \tau \right) \|x\|_\infty \\
&\leq \left(e^{\|A\|_\infty \tau} - 1 - \|A\|_\infty \tau \right) \sup_{x \in Z} \|x\|_\infty = \alpha_\tau . \qquad (7.18)
\end{aligned}
$$

Consider now the following set containing all the estimates of the form (7.17):

$$
C = \left\{ x' \in \mathbb{R}^n \mid x' = x + \frac{t}{\tau} \left(e^{tA} x - x \right) \text{ with } x \in Z \text{ and } t \in [0, \tau] \right\}.
$$

Since $t \in [0, \tau]$ implies $t/\tau \in [0, 1]$ we see that $C \subseteq Z_a$. Therefore, by inflating Z_a by the error in (7.18) we conclude that $\mathcal{R}_{[0,\tau]}(Z) \subseteq Z_a + \mathcal{C}_{\alpha_\tau} = Z_b$ as desired. □

Given some time horizon $T \in \mathbb{R}^+$ and a desired discretization step $\Delta \in \mathbb{R}^+$ such that $T/\Delta = k \in \mathbb{N}$ we can decompose $\mathcal{R}_{[0,T]}(Z)$ as:

$$
\mathcal{R}_{[0,T]}(Z) = \mathcal{R}_{[0,\Delta]}(Z) \cup \mathcal{R}_{[\Delta,2\Delta]}(Z) \cup \ldots \cup \mathcal{R}_{[(k-1)\Delta,k\Delta]}(Z). \qquad (7.19)
$$

The first set in this union can be over-approximated by Z_b. Moreover, the remaining sets can also be recursively over-approximated using Z_b as we now show. Applying $e^{A\Delta}$ on both sides of $\mathcal{R}_{[0,\Delta]}(Z) = \cup_{t \in [0,\Delta]} e^{At}(Z)$ we obtain:

$$
e^{A\Delta}(\mathcal{R}_{[0,\Delta]}(Z)) = e^{A\Delta} \left(\bigcup_{t \in [0,\Delta]} e^{At}(Z) \right) = \bigcup_{t \in [\Delta,2\Delta]} e^{At}(Z) = \mathcal{R}_{[\Delta,2\Delta]}(Z)
$$

therefore $\mathcal{R}_{[0,\Delta]}(Z) \subseteq Z_b$ implies $\mathcal{R}_{[\Delta,2\Delta]}(Z) \subseteq e^{A\Delta}(Z_b)$. Based on (7.19) we can over-approximate $\mathcal{R}_{[0,T]}(Z)$ by:

$$
\mathcal{R}_{[0,T]}(Z) \subseteq Z_b \cup e^{A\Delta}(Z_b) \cup e^{A2\Delta}(Z_b) \cup \ldots \cup e^{Ak\Delta}(Z_b). \qquad (7.20)
$$

We note that approximation errors are only introduced in the computation of Z_b and these errors can be made arbitrarily small by reducing Δ. Therefore, the computation of the reachable set by equation (7.20) can be made arbitrarily precise by suitably reducing Δ.

Example 7.32. We revisit Example 7.15 to illustrate the computation of the reachable set. In Figure 7.19 we show the set of initial states L and $e^{A\Delta}(L)$ for $\Delta = 0.01$. The trajectories of the vertices are also represented. Since Z_a is not guaranteed to contain all the points in $\mathcal{R}_{[0,\tau]}(L)$, Z_a is inflated to Z_b according to (7.16) with $\alpha_\tau = 0.103304$. Both Z_a and Z_b are depicted in Figure 7.19

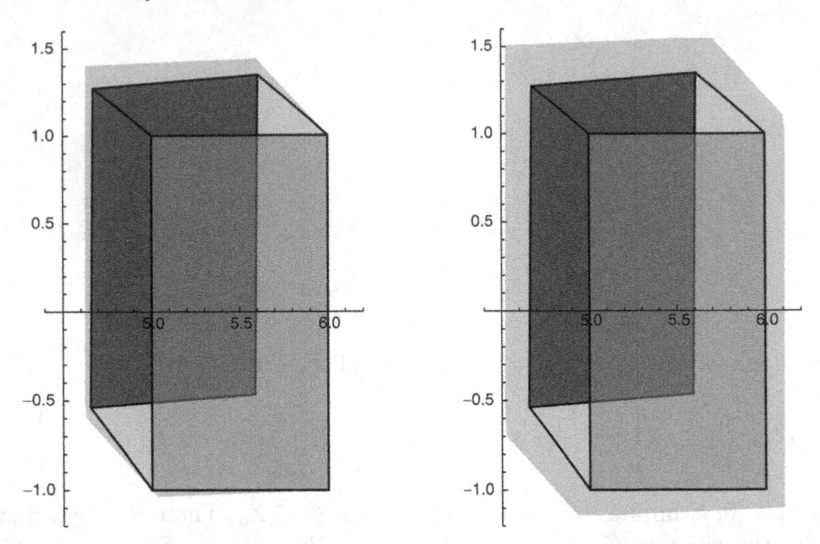

Fig. 7.19. Representation of the zonotopes: L, $e^{A\Delta}(L)$, Z_a, and Z_b. Set L is light-colored while $e^{A\Delta}(L)$ is dark-colored. The left figure displays also the zonotope Z_a in light gray while the right figure displays the zonotope Z_b in light gray.

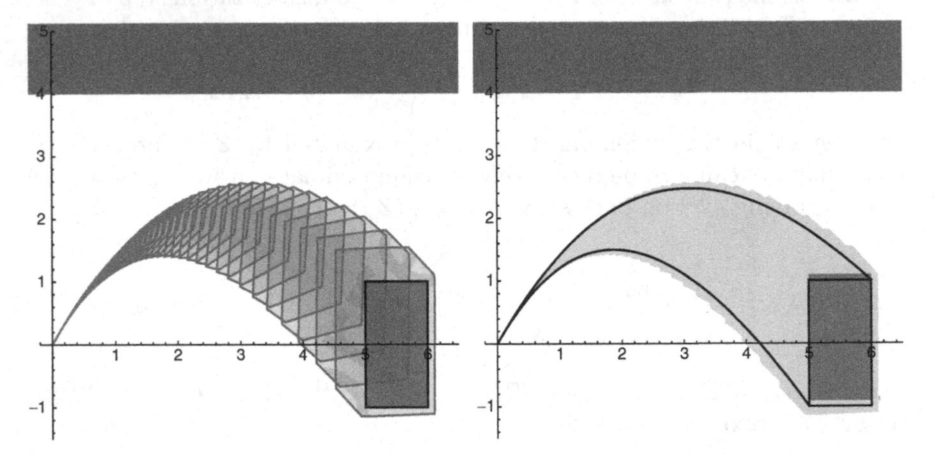

Fig. 7.20. Over-approximation of the reachable set. The left figure shows the intermediate zonotopes and the right figure shows the over-approximation of the reachable set in gray. Also shown are the true trajectories of the vertices delimiting the true reachable set. In both figures the set of unsafe states is dark-colored.

while in Figure 7.20 we show the over-approximation of the reachable set $\mathcal{R}_{[0,1]}(L)$. We can clearly see that the over-approximation does not intersect the set of unsafe states:

$$B = \{(x_1, x_2) \in \mathbb{R}^2 \mid x_2 \leq -4 \ \vee \ x_2 \geq 4)\}$$

from which we conclude safety. $\lhd$

7.7 Advanced topics

Several results in this chapter apply only to dynamical systems $\Sigma = (\mathbb{R}^m, f)$ where $f(x) = Ax$ is a linear map described by the matrix $A \in \mathbb{R}^{n \times n}$. In this section we describe a class on nonlinear differential equations that can be transformed into linear differential equations on a higher dimensional space. The following discussion requires some familiarity with more advanced mathematical concepts.

We start by relating dynamical systems, living on different spaces, by a map.

Definition 7.33. *Let $\Sigma_a = (\mathbb{R}^m, f)$ and $\Sigma_b = (\mathbb{R}^n, g)$ be dynamical systems and let $\phi : \mathbb{R}^m \to \mathbb{R}^n$ be a smooth map. Dynamical system Σ_a is said to be* ϕ-related *to dynamical system Σ_b if the following equality holds for every $x \in \mathbb{R}^m$:*

$$\frac{\partial \phi}{\partial x} f(x) = g \circ \phi(x). \tag{7.21}$$

Relating dynamical systems allows us to relate also their solutions.

Proposition 7.34. *Let $\Sigma_a = (\mathbb{R}^m, f)$ and $\Sigma_b = (\mathbb{R}^n, g)$ be dynamical systems and denote by ξ and ζ the solution of the differential equation defined by Σ_a and Σ_b, respectively. Dynamical system Σ_a is ϕ-related to dynamical system Σ_b iff the following equality holds for all $x \in \mathbb{R}^m$ and for all $t \in \mathbb{R}$ for which ξ and ζ are defined:*

$$\phi \circ \xi_x(t) = \zeta_{\phi(x)}(t). \tag{7.22}$$

When Σ_a is ϕ-related to Σ_b and Σ_b is "simpler" than Σ_a, equality (7.22) tell us that instead of studying the solutions ξ with initial condition x we can study the solutions ζ of the simpler system Σ_b with initial condition $\phi(x)$. Furthermore, when ϕ is an injective map, no information is lost in this process. We now identify a class of dynamical systems that can be ϕ-related to dynamical systems defining linear differential equations.

We introduce the main idea through a scalar differential equation $\dot{\xi} = f(\xi)$, $f : \mathbb{R} \to \mathbb{R}$. The key assumption we make is *local finiteness*. We say that $(\mathbb{R}, f)$ is a locally finite dynamical system if there exist a natural number k and a sequence of real numbers $\alpha_0, \alpha_1, \ldots, \alpha_{k-1}$ such that:

$$L_f^k 1_\mathbb{R} = \alpha_0 1_\mathbb{R} + \alpha_1 L_f 1_\mathbb{R} + \alpha_2 L_f^2 1_\mathbb{R} + \ldots + \alpha_{k-1} L_f^{k-1} 1_\mathbb{R}. \tag{7.23}$$

Under the local finiteness assumption, we can construct a linear differential equation $\zeta = g(\zeta) = A\zeta$ defined by the A matrix:

$$A = \begin{bmatrix} 0 & 1 & 0 & \ldots & 0 & 0 \\ 0 & 0 & 1 & \ldots & 0 & 0 \\ \vdots & & & \ddots & & \vdots \\ 0 & 0 & 0 & \ldots & 1 & 0 \\ 0 & 0 & 0 & \ldots & 0 & 1 \\ \alpha_0 & \alpha_1 & \alpha_2 & \ldots & \alpha_{k-2} & \alpha_{k-1} \end{bmatrix}$$

and we can construct the map $\phi : \mathbb{R} \to \mathbb{R}^k$ defined by:

$$\phi_1 = 1_{\mathbb{R}}, \quad \phi_2 = L_f 1_{\mathbb{R}}, \quad \phi_3 = L_f^2 1_{\mathbb{R}}, \quad \ldots \quad \phi_k = L_f^{k-1} 1_{\mathbb{R}}.$$

A simple computation now shows that (7.21) holds:

$$\frac{\partial \phi}{\partial x} f(x) = \begin{bmatrix} L_f 1_{\mathbb{R}}(x) \\ L_f^2 1_{\mathbb{R}}(x) \\ L_f^3 1_{\mathbb{R}}(x) \\ \ldots \\ L_f^k 1_{\mathbb{R}}(x) \end{bmatrix} = A\phi(x)$$

so that $\Sigma_a = (\mathbb{R}, f)$ is ϕ-related to $\Sigma_b = (\mathbb{R}^k, g)$ with $g(y) = Ay$. This process can be generalized to differential equations on $\mathbb{R}^m$ by replacing $1_{\mathbb{R}}$ with $1_{\mathbb{R}^m}$. In order to summarize the previous discussion, we introduce the concept of locally finite dynamical system.

Definition 7.35 (Locally finite dynamical system). *A dynamical system $\Sigma = (\mathbb{R}^m, f)$ is said to be* locally finite *if for every projection map $\pi_i : \mathbb{R}^m \to \mathbb{R}$, $i = 1, 2, \ldots, m$, the $\mathbb{R}$-vector space spanned by the functions:*

$$\pi_i, L_f(\pi_i), L_f^2(\pi_i), L_f^3(\pi_i), \ldots$$

is finite dimensional.

When the vector space described in the previous definition is finite dimensional, say of dimension k, we can write $L_f^k(\pi_i)$ as a linear combination of $L_f^j(\pi_i)$ for $j = 0, \ldots, k-1$. This is a generalization of the requirement (7.23) to the case $m > 1$. The map ϕ can be constructed in the same manner by starting with $\phi_1 = 1_{\mathbb{R}^m}$ and the construction of the A matrix follows from the specific coefficients appearing in the linear combinations defining the m scalar functions $L_f^k \pi_i$ for $i = 1, \ldots, m$.

Theorem 7.36. *Let $\Sigma_a = (\mathbb{R}^m, f)$ be a locally finite dynamical system. Then, there exists an injective and smooth map $\phi : \mathbb{R}^m \to \mathbb{R}^n$, with $n \geq m$, and there exists a dynamical system $\Sigma_b = (\mathbb{R}^n, g)$, with $g(y) = Ay$ a linear map, such that Σ_a is ϕ-related to Σ_b.*

7.8 Notes

In the literature, several different notions and mathematical formalizations of hybrid systems coexist. The one presented in this chapter represents a reasonable compromise between generality and tractability. It was inspired in the hybrid automaton model in [Hen96].

Timed automata were introduced by Alur and Dill in [AD90, AD94] and since then a wealth of deep results on verification and control appeared in the literature. We limited the discussion of timed automata to the rudiments of quotient based abstractions that set tone for the whole chapter. Introductory expositions on timed automata include Chapter 17 in [CGP99] and [Alu99, AM04, BY04].

Order-minimality was first used to show existence of symbolic models for hybrid systems in [LPS99, LPS00]. A much more insightful proof of a slight generalization of these results appears in [BM05]. The proof of Theorem 7.13 was based on [BM05]. The early work on decidability of several verification problems for restricted classes of hybrid systems, reported in [ACH+95, HKPV98], can now be understood as a consequence of Lemma 7.7 and Theorem 7.13. The proof of Theorem 7.11 can be found in [vdD98].

Sign based abstractions are discussed in [Tiw08] where the hybrid version of Proposition 7.21 is stated and proved. The reader can also find in this reference complete proofs for Proposition 7.21 and Theorem 7.25. Other abstraction techniques based on the analysis of Lie derivatives include [SSM06, PC08].

Barrier certificates for safety verification were introduced in [PR07, PJP07] where it is proved, under mild additional assumptions, a converse to Theorem 7.27: safety implies the existence of a barrier certificate. Although there is no systematic way of constructing barrier certificates, they can be searched for using an efficient convex programming technique called "sums of squares" [PPP02].

There is a large literature on reachability computation using different set representations such as a ellipsoids [KV00, BT00], polyhedra [ABDM00, CK03], level-sets [TMBO03], etc. The results in Section 7.6 are from [Gir05] and represent a very particular view of reachability computation.

Locally finite dynamical systems are discussed in [vdE94] under the name of locally finite derivations. In this reference the reader can find a proof of Theorem 7.36. This result can also be seen as a special case of similar results for control systems reported in [LM86]. The proof of Proposition 7.34 can be found, among other sources, in [AMR88].

8

Exact symbolic models for control

The prevalent role that software plays in modern complex engineering systems creates new control problems combining requirements of finite-state and of infinite-state nature. The existence of symbolic models for control systems suggests that we can use these abstractions to synthesize controllers enforcing finite-state requirements while accounting for the infinite-state dynamics. In this chapter, we discuss two classes of control systems admitting finite-state abstractions and how these can be used for control. Although the starting point is an infinite-state control system, the synthesized controllers are hybrid since they enforce finite-state and infinite-state requirements.

Notation

Given an equivalence relation Q on a set Z we denote by $[z]$ the equivalence class of $z \in Z$, by Z/Q the set of all equivalence classes, and by $\pi_Q : Z \to Z/Q$ the natural projection map taking a point $z \in Z$ to its equivalence class $\pi(z) = [z] \in Z/Q$. We say that an equivalence relation is finite when it has finitely many equivalence classes. An equivalence relation Q refines an equivalence relation R when $(z, z') \in Q$ implies $(z, z') \in R$. A finite partition $\mathcal{P}$ of a set Z is a finite collection of sets $\mathcal{P} = \{P_i\}_{i \in I}$ satisfying $P_i \subseteq Z$, $\cup_{i \in I} P_i = Z$, and for any $i, j \in I$, $i \neq j$ implies $P_i \cap P_j = \varnothing$. A partition $\mathcal{P}$ of a set Z defines an equivalence relation P on Z by declaring each set $P_i \in \mathcal{P}$ to be an equivalence class of P.

The transpose of a vector $x \in \mathbb{R}^n$ is denoted by x^T while the ith component is denoted by x_i. The topological closure of a set $Z \subseteq \mathbb{R}^n$ is denoted by $\overline{Z}$ and the interior of Z is denoted by int Z.

A function $f :]a, b[\to \mathbb{R}^n$, $a, b \in \mathbb{R}$, is said to be piecewise continuous or piecewise continuously differentiable if there exists an ordered sequence of real numbers $a = i_1 < i_2 < \ldots < i_k = b$ such that for every $j \in \{1, 2, \ldots, k-1\}$, the restriction of f to the interval $]i_j, i_{j+1}[$ is continuous or has continuous

P. Tabuada, *Verification and Control of Hybrid Systems: A Symbolic Approach*,
DOI: 10.1007/978-1-4419-0224-5_8, © Springer Science + Business Media, LLC 2009

derivative, respectively. A piecewise continuous function $f :]a, b[\rightarrow \mathbb{R}^n$ is essentially bounded if there exists a compact set $K \subset \mathbb{R}^n$ such that $f(t) \in K$ for almost all $t \in]a, b[$.

8.1 Control systems as systems

The notion of system introduced in Chapter 1 can be used to model control systems in discrete-time and continuous-time.

8.1.1 Discrete-time control systems

A discrete-time control system consists of a smooth map $f : \mathbb{R}^n \times \mathbb{R}^m \rightarrow \mathbb{R}^n$ describing the state $f(x, u) \in \mathbb{R}^n$ that is reached by applying the input $u \in \mathbb{R}^m$ at the state $x \in \mathbb{R}^n$. The controlled economy model in Chapter 1 offers an example of a discrete-time control system.

Definition 8.1 (Discrete-time control system). *A* discrete-time control system *is a triple* $\Sigma = (\mathbb{R}^n, \mathbb{R}^m, f)$ *consisting of:*

- *the state space* $\mathbb{R}^n$*;*
- *the input space* $\mathbb{R}^m$*;*
- *a smooth map* $f : \mathbb{R}^n \times \mathbb{R}^m \rightarrow \mathbb{R}^n$*.*

Using an equivalence relation on $\mathbb{R}^n$ to define the outputs, we can describe discrete-time control systems using the notion of system adopted in this book.

Definition 8.2. *Let* $\Sigma = (\mathbb{R}^n, \mathbb{R}^m, f)$ *be a discrete-time control system and let* Q *be an equivalence relation on* $\mathbb{R}^n$*. The system associated with* Σ *and* Q*, denoted by* $S_Q(\Sigma)$*, consists of:*

- $X = \mathbb{R}^n$*;*
- $U = \mathbb{R}^m$*;*
- $x \xrightarrow{u} x'$ *if* $x' = f(x, u)$*;*
- $Y = X/Q$*;*
- $H = \pi_Q$*.*

In control theory, and also in this chapter, the notion of controllability plays a fundamental role.

Definition 8.3 (Controllable discrete-time system). *A discrete-time control system* Σ *is said to be* controllable *if for any two states* $x, x' \in \mathbb{R}^n$ *there exists a finite internal behavior of* $S_Q(\Sigma)$*:*

$$x_0 \xrightarrow{u_0} x_1 \xrightarrow{u_1} x_2 \xrightarrow{u_2} \ldots \xrightarrow{u_{k-1}} x_k$$

with $x_0 = x$ *and* $x_k = x'$*.*

Controllability admits a very simple characterization for linear control systems. Recall that a discrete-time control system is said to be linear when:

$$f(x, u) = Ax + Bu$$

for matrices $A \in \mathbb{R}^{n \times n}$ and $B \in \mathbb{R}^{n \times m}$. We denote a linear control system by the quadruple $\Sigma = (\mathbb{R}^n, \mathbb{R}^m, A, B)$. The following classical result in linear systems theory characterizes controllability by a rank condition.

Theorem 8.4. *A discrete-time linear control system $\Sigma = (\mathbb{R}^n, \mathbb{R}^m, A, B)$ is controllable iff* $\operatorname{rank} \mathcal{C} = n$ *where $\mathcal{C}$ is the controllability matrix of Σ defined by:*

$$\mathcal{C} = \left[B | AB | A^2 B | \ldots | A^{n-1} B\right].$$

8.1.2 Continuous-time control systems

A control system in continuous-time is described by a differential equation $\dot{\xi} = f(\xi, \upsilon)$ whose solution ξ can be influenced by external inputs υ. An example is provided by the satellite equipped with gas jets, discussed in Chapter 1.

Definition 8.5 (Continuous-time control system). *A* continuous-time control system *is a triple $\Sigma = (\mathbb{R}^n, \mathcal{U}, f)$ consisting of:*

- *the state space $\mathbb{R}^n$;*
- *a set of input curves $\mathcal{U}$ whose elements are*[1] *essentially bounded piecewise continuous functions of time from intervals of the form $]a, b[\subseteq \mathbb{R}$ to $\mathbb{R}^m \subseteq \mathbb{R}^m$ with $a < 0 < b$;*
- *a smooth map $f : \mathbb{R}^n \times \mathbb{R}^m \to \mathbb{R}^n$.*

A piecewise continuously differentiable curve $\xi :]a, b[\to \mathbb{R}^n$ is said to be a trajectory *or* solution *of Σ if there exist $\upsilon \in \mathcal{U}$ satisfying:*

$$\frac{d}{dt}\xi = f(\xi, \upsilon)$$

for almost all $t \in]a, b[$.

Although we have defined trajectories over open sets, we shall refer to trajectories $\xi : [0, \tau] \to \mathbb{R}^n$ defined on closed sets $[0, \tau]$, $\tau \in \mathbb{R}^+$, with the understanding of the existence of a trajectory $\xi' :]a, b[\to \mathbb{R}^n$ such that $\xi = \xi'|_{[0,\tau]}$. We also write $\xi_{x\upsilon}(t)$ to denote the point reached at time $t \in [0, \tau]$ under the

[1] We do not allow $\mathcal{U}$ to be the set of all curves from $]a, b[$ to $\mathbb{R}^m$. From a mathematical point of view, we restrict $\mathcal{U}$ to a class of functions that are regular enough to guarantee existence and uniqueness of solutions for the differential equation $\dot{\xi} = f(\xi, \upsilon)$. From an engineering point of view, the curves in $\mathcal{U}$ describe signals that are implemented by physical actuators. Hence, actuator technology poses a different set of limitations on the curves in $\mathcal{U}$.

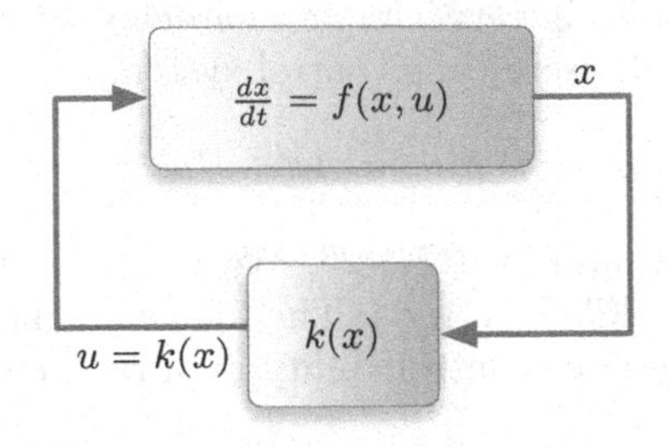

Fig. 8.1. Diagrammatic representation of a closed-loop system.

input υ from initial condition x. This point is uniquely determined since the assumptions on f and $\mathcal{U}$ ensure existence and uniqueness of trajectories.

A feedback law for a control system is a smooth map $k : \mathbb{R}^n \to \mathbb{R}^m$ transforming states $x \in \mathbb{R}^n$ into inputs $k(x) = u \in \mathbb{R}^m$. The term feedback describes the process of *feeding* the current state x *back* as the input $u = k(x)$. This process is diagrammatically illustrated in Figure 8.1 and results in a dynamical system, termed closed-loop system, defined by the differential equation:

$$\frac{d\xi}{dt} = f(\xi, k(\xi)).$$

The construction of a system $S_Q(\Sigma)$ from a dynamical system Σ and a finite equivalence relation, discussed in Chapter 7, can be generalized to continuous-time control systems.

Definition 8.6. *Let $\Sigma = (\mathbb{R}^n, \mathcal{U}, f)$ be a control system and let Q be an equivalence relation on $\mathbb{R}^n$. The system associated with Σ and Q, denoted by $S_Q(\Sigma)$, consists of:*

- $X = \mathbb{R}^n$;
- $U = \mathcal{U}$;
- $x \xrightarrow{\upsilon} x'$ *if any of the following two conditions is satisfied:*
  1. $\pi_Q(x) \neq \pi_Q(x')$, $\xi_{x\upsilon} : [0, \tau] \to \mathbb{R}^n$ *is a solution of* Σ *satisfying* $\xi_{x\upsilon}(\tau) = x'$, *and there exists* $\varepsilon \in [0, \tau]$ *satisfying one of the following:*
     a) $\pi_Q(\xi_{x\upsilon}(t)) = \pi_Q(x)$ *for* $t \in [0, \varepsilon[$, $\pi_Q(\xi_{x\upsilon}(t)) = \pi_Q(x')$ *for* $t \in [\varepsilon, \tau]$, *and* $\pi_Q(x) \neq \pi_Q(x')$;
     b) $\pi_Q(\xi_{x\upsilon}(t)) = \pi_Q(x)$ *for* $t \in [0, \varepsilon]$, $\pi_Q(\xi_{x\upsilon}(t)) = \pi_Q(x')$ *for* $t \in]\varepsilon, \tau]$, *and* $\pi_Q(x) \neq \pi_Q(x')$;
  2. $\pi_Q(x) = \pi_Q(x')$ *and* $\xi_{x\upsilon} : \mathbb{R}_0^+ \to \mathbb{R}^n$ *is a solution of* Σ *satisfying* $\xi_{x\upsilon}(\tau) = x'$ *and* $\pi_Q(\xi_{x\upsilon}(t)) = \pi_Q(x)$ *for all* $t \in \mathbb{R}_0^+$;
- $Y = X/Q$;
- $H = \pi_Q$.

Contrary to dynamical systems, we assume the set of initial states X_0 to equal the set of states X since in the vast majority of control problems we do not have the possibility of initializing the state.

8.2 Controller refinement

The synthesis methods discussed in Chapter 7 produce a controller based on a finite-state model of the system to be controlled. When this model is an abstraction S_{abs} of an infinite-state system S, it is natural to ask how to refine the controller synthesized for S_{abs} so that it can be applied to S. For safety verification problems, the crucial relationship between a model S and its abstraction S_{abs} is that S_{abs} simulates S. When this holds, a positive answer to the safety verification problem for S_{abs} implies a positive answer to the safety verification problem for S. For control problems, a similar relationship holds provided that we work with alternating simulation relations.

Proposition 8.7. *Let S_a, S_b, and S_c be systems with the same output set, assume that S_c is feedback composable with S_a, and let ${}_cR_a$ be the corresponding alternating simulation relation. If there exists an alternating simulation relation ${}_aR_b$ from S_a to S_b, then $S_c \times_{{}_cR_a^e} S_a$ is feedback composable with S_b and the corresponding alternating simulation relation is given by:*

$$ {}_{ca}R_b = \{((x_c, x_a), x_b) \in (X_c \times X_a) \times X_b \mid (x_c, x_a) \in {}_cR_a \wedge (x_a, x_b) \in {}_aR_b\}. $$

Proof. The proof consists in showing that the relation ${}_{ca}R_b$ satisfies all the requirements in Definition 4.19.

We start with the first requirement. Let $(x_{c0}, x_{a0}) \in X_{ca0}$. By definition of feedback composition $(x_{c0}, x_{a0}) \in {}_cR_a$ and $x_{a0} \in X_{a0}$. Invoking now the alternating simulation relation ${}_aR_b$ from S_a to S_b, there exists $x_{b0} \in X_{b0}$ satisfying $(x_{a0}, x_{b0}) \in {}_aR_b$. Consequently, $((x_{c0}, x_{a0}), x_{b0}) \in {}_{ca}R_b$.

The second requirement follows immediately from the definition of ${}_{ca}R_b$.

Consider now the third requirement, let $((x_c, x_a), x_b) \in {}_{ca}R_b$, and let (u_c, u_a) be any element of $U_{ca}(x_c, x_a)$. By definition of feedback composition, and since ${}_cR_a$ is an alternating simulation relation, we know that:

$$ \forall x_a' \in \mathrm{Post}_{u_a}(x_a) \ \exists x_c' \in \mathrm{Post}_{u_c}(x_c) \quad (x_c', x_a') \in {}_cR_a. \tag{8.1} $$

Moreover, as ${}_aR_b$ is an alternating simulation relation, we also have:

$$ \exists u_b \in U_b(x_b) \ \forall x_b' \in \mathrm{Post}_{u_b}(x_b) \ \exists x_a' \in \mathrm{Post}_{u_a}(x_a) \quad (x_a, x_b) \in {}_aR_b. \tag{8.2} $$

Combining (8.1) with (8.2) we conclude the existence of $u_b \in U_b(x_b)$ such that for all $x_b' \in \mathrm{Post}_{u_b}(x_b)$ there exists $(x_c', x_a') \in \mathrm{Post}_{(u_c,u_a)}(x_c, x_a)$ for which $((x_c', x_a'), x_b') \in {}_{ca}R_b$ thus concluding the proof. □

Proposition 8.7 can be summarized by the following implication:

$$S_c \preceq_{\mathcal{AS}} S_a \wedge S_a \preceq_{\mathcal{AS}} S_b \implies S_c \times_{\mathcal{F}} S_a \preceq_{\mathcal{AS}} S_b \tag{8.3}$$

and used to explain how a controller S_{cont} synthesized for an abstraction S_{abs} of a system S can be refined to a controller for S. If S_{cont} solves a simulation game for system S_{abs} and specification S_{spec}, we have:

$$S_{cont} \preceq_{\mathcal{AS}} S_{abs}, \tag{8.4}$$

$$S_{cont} \times_{\mathcal{F}} S_{abs} \preceq_{\mathcal{S}} S_{spec}. \tag{8.5}$$

Moreover, if S_{abs} is an abstraction of S in the sense:

$$S_{abs} \preceq_{\mathcal{AS}} S, \tag{8.6}$$

we can use Proposition 8.7, or implication (8.3), applied to the conjunction of (8.4) with (8.6), to conclude that $S_{cont} \times_{\mathcal{F}} S_{abs}$ is feedback composable with S. This suggests that we use the controller:

$$S'_{cont} = S_{cont} \times_{\mathcal{F}} S_{abs}$$

to control S. Indeed:

$$S'_{cont} \times_{\mathcal{G}} S = (S_{cont} \times_{\mathcal{F}} S_{abs}) \times_{\mathcal{G}} S \preceq_{\mathcal{S}} S_{cont} \times_{\mathcal{F}} S_{abs} \preceq_{\mathcal{S}} S_{spec}$$

where the first simulation relation follows from Proposition 6.3 and the second follows from (8.5). We conclude that controller $S'_{cont} = S_{cont} \times_{\mathcal{F}} S_{abs}$ enforces the specification S_{spec} on the original system S. However, if no controller exists for S_{abs}, we cannot conclude the nonexistence of a controller for S. A stronger claim can be made when the abstraction S_{abs} is related to the original model S by an alternating bisimulation relation. In such case, existence of a controller S'_{cont} enforcing the specification S_{spec} on S implies, by Proposition 8.7, that $S'_{cont} \times_{\mathcal{F}} S$ is a controller enforcing S_{spec} on S_{abs}. Hence, when the abstract model S_{abs} is alternatingly bisimilar to the original model S, a controller exists for S_{abs} iff a controller exists for S.

8.3 Discrete-time linear control systems

We saw in Section 7.3 that by placing certain restrictions on the eigenvalues of linear dynamical systems, finite-state bisimilar abstractions are guaranteed to exist. In this section we follow a complementary approach: instead of constraining the dynamics, we constrain the outputs. Intuitively, we work with partitions consisting of sets that are adapted to the dynamics of the control system.

Definition 8.8 (Adapted sets). *Let $\Sigma = (\mathbb{R}^n, \mathbb{R}^m, A, B)$ be a discrete-time linear system and consider a collection of m vectors $c_1, c_2, \ldots, c_m \in \mathbb{R}^n$ for which there exist numbers $\nu_1, \nu_2, \ldots, \nu_m \in \mathbb{N}$ satisfying:*

1. $c_r^T b_r = 0,\ c_r^T A b_r = 0, \ldots, c_r^T A^{\nu_r-2} b_r = 0,\ c_r^T A^{\nu_r-1} b_r \neq 0,\ r = 1, \ldots, m;$
2. *the vectors $c_1^T A^{\nu_1} B, c_2^T A^{\nu_2} B, \ldots, c_m^T A^{\nu_m} B$ are linearly independent,*

where $b_1, b_2, \ldots, b_m$ are the columns of B. The class of subsets of $\mathbb{R}^n$ adapted *to Σ is formed by finite unions of sets defined by conjunctions of conditions of the form $f \sim 0$ with $f = \pm c_r^T A^l x \pm e$, $e \in \mathbb{R}$, $l \in \{0, 1, \ldots, \nu_r - 1\}$, and $\sim \in \{=, >\}$.*

Example 8.9. We revisit the controlled model for the national income, discussed in Chapter 1 and repeated here for convenience:

$$\begin{aligned} c(n+1) &= \alpha\big(c(n) + i(n) + g(n)\big) \\ i(n+1) &= \beta\alpha\big(c(n) + i(n) + g(n)\big) - \beta c(n) \\ g(n+1) &= d(n). \end{aligned}$$

Taking $\alpha = \frac{1}{2}$ and $\beta = 2$, the corresponding A and B matrices are given by:

$$A = \begin{bmatrix} \frac{1}{2} & \frac{1}{2} & \frac{1}{2} \\ -1 & 1 & 1 \\ 0 & 0 & 0 \end{bmatrix} \qquad B = \begin{bmatrix} 0 \\ 0 \\ 1 \end{bmatrix}. \tag{8.7}$$

For this system $m = 1$ and adapted sets are defined using a single vector c_1. The choice:

$$c_1 = \begin{bmatrix} 1 \\ 0 \\ 0 \end{bmatrix} \tag{8.8}$$

satisfies $c_1^T B = 0$, $c_1^T AB \neq 0$ and thus $\nu_1 = 2$. Note that in this case the second condition in Definition 8.8 is trivially satisfied since $m = 1$. Noting that $c_1^T x = c$ and $c_1^T Ax = \frac{1}{2}(c + i + g)$, the 8 possible functions f for a fixed $e \in \mathbb{R}$ are given by:

$$\begin{aligned} f_1(c, i, g) &= c + e, \\ f_2(c, i, g) &= c - e, \\ f_3(c, i, g) &= -c + e, \\ f_4(c, i, g) &= -c - e, \\ f_5(c, i, g) &= \frac{1}{2}(c + i + g) + e, \\ f_6(c, i, g) &= \frac{1}{2}(c + i + g) - e, \\ f_7(c, i, g) &= -\frac{1}{2}(c + i + g) + e, \\ f_8(c, i, g) &= -\frac{1}{2}(c + i + g) - e. \end{aligned} \quad \triangleleft$$

Adapted sets exist for any discrete-time linear system Σ. It suffices to take $\nu_1 = \nu_2 = \ldots = \nu_m = 1$ and to define the vectors c_i by $c_i = b_i$. More interesting choices for the vectors c_i are discussed once we show how finite partitions defined by adapted sets, termed adapted finite partitions, lead to finite-state bisimilar systems.

Theorem 8.10. *Let $\Sigma = (\mathbb{R}^n, \mathbb{R}^m, f)$ be a discrete-time linear system. For any finite partition $\mathcal{P}$ of $\mathbb{R}^n$ adapted to Σ there exists a finite-state system bisimilar to $S_Q(\Sigma)$ where Q is the equivalence relation defined by $\mathcal{P}$.*

Proof. The first step in the proof consists in making a change of input coordinates. We define new inputs v by $u = Lv$ where $L \in \mathbb{R}^{m\times m}$ is defined by:

$$DL = I_m \tag{8.9}$$

with I_m the $m \times m$ identity matrix and D the matrix having $c_i^T A^{\nu_i - 1}B$ as its i-th row. Matrix L exists and is invertible since, by assumption, the vectors $c_i^T A^{\nu_i-1}B$ are linearly independent. Therefore, if we denote by Σ' the control system $(\mathbb{R}^n, \mathbb{R}^m, A, B')$ with $B' = BL$, it follows that $S_Q(\Sigma)$ is bisimilar to $S_Q(\Sigma')$. This can be seen by using the identity relation on $\mathbb{R}^n$ as bisimulation relation since $x \xrightarrow{u} x'$ in $S_Q(\Sigma)$ iff $x \xrightarrow{L^{-1}u} x'$ in $S_Q(\Sigma')$. Therefore, in the remaining proof we work with Σ'. We shall not prove it here, but if follows from well established results in systems theory [AM97] that for every $l < \nu_r - 1$ we have:

$$c_r^T A^l B = 0. \tag{8.10}$$

Every set P in $\mathcal{P}$ is a finite union of sets defined by conditions of the form $f \sim 0$ with $f = \pm c_r^T A^l x \pm e$ and $\sim \in \{=, >\}$. We now construct a new set of functions consisting of all the functions $g(x) = \pm c_r^T A^l x \pm e$ in which $r \in \{1, 2, \ldots, m\}$, $0 \le l \le \nu_r - 1$, and the constants e appear in the functions f defining the sets $P \in \mathcal{P}$. The functions g induce an equivalence relation R on $\mathbb{R}^n$ defined by $(x, x') \in R$ if $\text{sign} \circ g(x) = \text{sign} \circ g(x')$ for every function g. Note that R is finite as we have finitely many functions g. We claim that R is a bisimulation relation between $S_Q(\Sigma)$ and $S_Q(\Sigma)$. Note that for any $x \in \mathbb{R}^n$, $(x, x) \in R$ which implies that conditions 1a and 1b in Definition 4.13 are satisfied. Moreover, $(x, x') \in R$ implies $\pi_Q(x) = \pi_Q(x')$ since every function f is also a function g, that is, R refines the equivalence relation defined by $\mathcal{P}$. Hence, condition 2 in Definition 4.13 is also satisfied. We now focus on condition 3. Let $(x, x') \in R$ and assume that $x \xrightarrow{u} x''$, or equivalently, that $x'' = Ax + BLv$. We claim that $x''' = Ax' + BLv'$ satisfies $(x'', x''') \in R$ where:

$$v_i' = c_i^T A^{\nu_i}(x - x') + v_i. \tag{8.11}$$

Let g be any function used to construct R. Noting that:

$$g(x''') = \pm c_r^T A^l x''' \pm e = \pm c_r^T A^l (Ax' + BLv') \pm e = \pm c_r^T A^{l+1} x' \pm c_r^T A^l BLv' \pm e$$

we have the following two situations. When $l < \nu_r - 1$, $c_r^T A^l B L = 0$, by (8.10), and $\text{sign} \circ g(x'') = \text{sign} \circ g(x''')$ becomes:

$$\text{sign}(\pm c_r^T A^{l+1} x \pm e) = \text{sign}(\pm c_r^T A^{l+1} x' \pm e)$$

which holds by definition of R and by the membership $(x, x') \in R$. If $l = \nu_r - 1$, then $g(x''') = \pm c_r^T A^{\nu_r} x' \pm c_r^T A^{\nu_r - 1} B L v' \pm e$ with $c_r^T A^{\nu_r - 1} B L \neq 0$. Moreover:

$$\begin{aligned} g(x''') &= \pm c_r^T A^{\nu_r} x' \pm c_r^T A^{\nu_r - 1} B L v' \pm e \\ &= \pm c_r^T A^{\nu_r} x' \pm c_r^T A^{\nu_r} (x - x') \pm c_r^T A^{\nu_r - 1} B L v \pm e \quad \text{by (8.11) and (8.9)} \\ &= \pm c_r^T A^{\nu_r} x \pm c_r^T A^{\nu_r - 1} B L v \pm e = g(x''). \end{aligned}$$

It then follows that $\text{sign} \circ g(x''') = \text{sign} \circ g(x'')$ for any g used to construct R and we conclude $(x'', x''') \in R$. By symmetry, condition 3b in Definition 4.13 also holds and the result now follows from Theorem 4.18. □

The notion of quotient system, introduced in Definition 4.17, preserves determinism in the sense that the quotient of a deterministic system is a deterministic system. Hence, $S_Q(\Sigma)$ being deterministic, has a symbolic deterministic quotient S. We can thus refine controllers synthesized for S to controllers acting on $S_Q(\Sigma)$, as described in Section 8.2, since alternating simulation degenerates into simulation for deterministic systems.

The equivalence relation R used in the proof of Theorem 8.10 is not guaranteed to be the maximal bisimulation relation between $S_Q(\Sigma)$ and $S_Q(\Sigma)$. Therefore, the resulting finite-state system S, constructed as the quotient $S = (S_Q(\Sigma))_{/R}$, is not guaranteed to have the smallest number of states. Although we could use the operator G, introduced in Chapter 5, to construct the maximal bisimulation relation, we would not be exploiting the fact that we seek a bisimulation relation between $S_Q(\Sigma)$ and itself. Using the operator $\text{Pre} : 2^X \to 2^X$ defined, for any $W \subseteq X$, by:

$$\text{Pre}(W) = \{x \in X \mid x \xrightarrow{u} x' \text{ for some } u \in U \text{ and } x' \in W\}, \qquad (8.12)$$

and the fact that we are dealing with deterministic systems, we can specialize the algorithm defined by G. Starting from a partition $\mathcal{P}$ we can refine it using the Pre operator until it stabilizes, *i.e.*, until $\text{Pre}(P)$ can be written as a union of sets in $\mathcal{P}$ for any set $P \in \mathcal{P}$. Once stabilization is achieved, the resulting partition defines a bisimulation relation between $S_Q(\Sigma)$ and $S_Q(\Sigma)$ since $P' \cap \text{Pre}(P) \neq \varnothing$ implies that every point $x \in P'$ also belongs to $\text{Pre}(P)$. These ideas lead to Algorithm 8.1 whose correctness can be easily proved using the fixed-point techniques of Chapter 5. Moreover, a careful analysis of the proof of Theorem 8.10 reveals that the number of equivalence classes in the equivalence relation computed by Algorithm 8.1 is $|\mathcal{P}|^\nu \leq |\mathcal{P}|^n$. In this expression, $|\mathcal{P}|$ denotes the number of elements in the partition $\mathcal{P}$ and ν is the largest number in the sequence $\nu_1, \nu_2, \ldots, \nu_m$ introduced in Definition 8.8. Consequently, the number of states in the finite-state abstraction S, whose existence is asserted by Theorem 8.10, is bounded by $|\mathcal{P}|^\nu$.

Input: Partition $\mathcal{P}$ and system S
Output: $\mathcal{P}'$
$\mathcal{P}' := \mathcal{P}$;
while $\exists P, P' \in \mathcal{P}'$ *such that* $\varnothing \neq P' \cap \text{Pre}(P) \neq P'$ **do**
  $P_a := P' \cap \text{Pre}(P)$;
  $P_b := P' \backslash \text{Pre}(P)$;
  $\mathcal{P}' := (\mathcal{P}' \backslash \{P'\}) \cup \{P_a, P_b\}$;
end

Algorithm 8.1: Computation of the largest bisimulation relation, respecting partition $\mathcal{P}$, between S and S.

Example 8.11. Consider again the controlled model for the national income and suppose that we are interested in reducing the internal consumption from 10 units to 2 units while keeping the national income above 20 units. Using the vector c_1 in (8.8) we define the finite partition $\mathcal{P}$ adapted to Σ by:

$$\begin{aligned} P_1 &= \{(c,i,g) \in \mathbb{R}^3 \mid c_1^T Ax < 10\} \\ P_2 &= \{(c,i,g) \in \mathbb{R}^3 \mid c_1^T Ax \geq 10 \wedge c_1^T x \leq 2\} \\ P_3 &= \{(c,i,g) \in \mathbb{R}^3 \mid c_1^T Ax \geq 10 \wedge 2 < c_1^T x < 10\} \\ P_4 &= \{(c,i,g) \in \mathbb{R}^3 \mid c_1^T Ax \geq 10 \wedge c_1^T x \geq 10\}. \end{aligned}$$

In terms of these four regions, our objective can be formulated as the existence of a control strategy driving all the points in region P_4 to region P_2 without entering region P_1.

In order to obtain a bisimulation relation, we refine this partition using Algorithm 8.1. We first apply Pre to P_1 and obtain the set of points $x \in \mathbb{R}^3$ satisfying $c_1^T A^2 x + c_1^T ABu < 10$. Since $c_1^T AB \neq 0$ we conclude that $\text{Pre}(P_1) = \mathbb{R}^3$ as we can always find a $u \in \mathbb{R}$ such that $c_1^T A^2 x + c_1^T ABu < 10$ is satisfied. Applying Pre to P_2 results in the set defined by the points $x \in \mathbb{R}^3$ satisfying $c_1^T A^2 x + c_1^T ABu \geq 10$ and $c_1^T Ax + c_1^T Bu \leq 2$. As we saw before, the first condition is satisfied by every point in $\mathbb{R}^3$. The second condition is equivalent to $c_1^T Ax \leq 2$ since $c_1^T B = 0$. Note now that $P_1 \cap \text{Pre}(P_2) \neq \varnothing$ and $P_1 \cap \text{Pre}(P_2) \neq P_1$. We thus split the set P_1 into the sets P_{1a} and P_{1b} defined by:

$$\begin{aligned} P_{1a} &= \{(c,i,g) \in \mathbb{R}^3 \mid c_1^T Ax \leq 2\} \\ P_{1b} &= \{(c,i,g) \in \mathbb{R}^3 \mid 2 < c_1^T Ax < 10\}. \end{aligned}$$

We can easily check that the resulting partition:

$$\begin{aligned} P_{1a} &= \{(c,i,g) \in \mathbb{R}^3 \mid c_1^T Ax \leq 2\} \\ P_{1b} &= \{(c,i,g) \in \mathbb{R}^3 \mid 2 < c_1^T Ax < 10\} \\ P_2 &= \{(c,i,g) \in \mathbb{R}^3 \mid c_1^T Ax \geq 10 \wedge c_1^T x \leq 2\} \\ P_3 &= \{(c,i,g) \in \mathbb{R}^3 \mid c_1^T Ax \geq 10 \wedge 2 < c_1^T x < 10\} \\ P_4 &= \{(c,i,g) \in \mathbb{R}^3 \mid c_1^T Ax \geq 10 \wedge c_1^T x \geq 10\} \end{aligned}$$

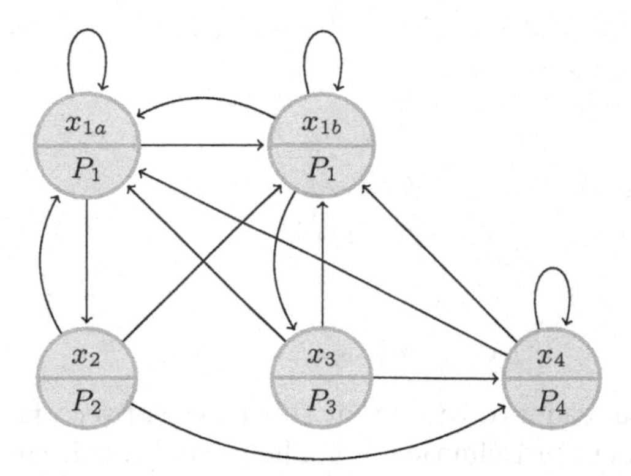

Fig. 8.2. Finite-state system $S = (S_Q(\Sigma))_{/R}$ bisimilar to $S_Q(\Sigma)$ with Σ defined by (8.7).

is already closed under the Pre operator and thus defines the maximal bisimulation relation R between $S_Q(\Sigma)$ and $S_Q(\Sigma)$. The resulting finite-state system $S = (S_Q(\Sigma))_{/R}$ is depicted in Figure 8.2 where each state x_i represents the set P_i. Inspecting Figure 8.2 we immediately see that it is not possible to reach P_2 from P_4 without passing through P_1. Since S is bisimilar to $S_Q(\Sigma)$ we conclude that the desired objective cannot be achieved in $S_Q(\Sigma)$ neither in S. When the finite-state abstraction is more complex, it is no longer possible to solve the control synthesis problem by inspection and one resorts of the fixed-point algorithms described in Chapter 6. $\triangleleft$

Theorem 8.10 critically depends on the use of adapted sets. Its practical importance is thus tied to the flexibility we have in using adapted sets to describe the problem being solved. A reasonable requirement is that we have n linearly independent vectors $c_r^T A^l$ at our disposal to define adapted sets. Although this is not guaranteed in general, n such linearly independent vectors do exist whenever Σ is controllable. In fact, existence of such vectors is equivalent to controllability. Furthermore, there exists a procedure to construct the vectors $c_1, \ldots, c_m$ that we now describe. We start by constructing an $n \times n$ matrix $\widetilde{\mathcal{C}}$ containing n linearly independent columns of $\mathcal{C}$. The columns of $\widetilde{\mathcal{C}}$ are found by sweeping the columns of $\mathcal{C}$ from left to right and extracting the first n linearly independent columns. Note that if the columns in B are linearly independent, then they are among the n linearly independent chosen columns. Once we identify these columns, we reorder them in the form:

$$\widetilde{\mathcal{C}} = \left[b_1|Ab_1|\ldots|A^{\nu_1-1}b_1|b_2|Ab_2|\ldots|A^{\nu_2-1}b_2|\ldots|b_m|Ab_m|\ldots|A^{\nu_m-1}b_m\right].$$

This reordering also defines the numbers $\nu_1, \nu_2, \ldots, \nu_m$ appearing in Definition 8.8. Since all the columns of $\widetilde{\mathcal{C}}$ are linearly independent, $\widetilde{\mathcal{C}}$ is invertible and its inverse satisfies:

$$\widetilde{\mathcal{C}}^{-1}\widetilde{\mathcal{C}} = I_n \tag{8.13}$$

where I_n is the $n \times n$ identity matrix. If we denote by q_i the ith row of $\widetilde{\mathcal{C}}^{-1}$, we define the vectors c_j as follows:

$$c_1^T = q_{\nu_1}, \tag{8.14}$$
$$c_2^T = q_{\nu_1+\nu_2}, \tag{8.15}$$
$$c_3^T = q_{\nu_1+\nu_2+\nu_3}, \tag{8.16}$$
$$\vdots \tag{8.17}$$
$$c_m^T = q_{\nu_1+\nu_2+\ldots+\nu_m} = q_n. \tag{8.18}$$

We leave it to the reader to verify that this choice of vectors and (8.13) imply the first requirement in Definition 8.8. The second requirement is satisfied by construction since all the rows in $\widetilde{\mathcal{C}}^{-1}$ are linearly independent.

Example 8.12. We illustrate the preceding construction using the matrices:

$$A = \begin{bmatrix} 1 & 5 & 3 & 4 \\ -2 & 4 & -1 & 3 \\ 0 & 0 & 3 & 1 \\ 2 & -1 & 0 & 1 \end{bmatrix} \qquad B = \begin{bmatrix} 1 & 3 \\ 0 & 2 \\ 2 & -4 \\ -1 & 1 \end{bmatrix}.$$

The controllability matrix is given by:

$$\mathcal{C} = \left[B|AB|A^2B\right] = \begin{bmatrix} 1 & 3 & 3 & 5 & -13 & 37 \\ 0 & 2 & -7 & 9 & -36 & 52 \\ 2 & -4 & 5 & -11 & 16 & -28 \\ -1 & 1 & 1 & 5 & 14 & 6 \end{bmatrix}$$

and the matrix $\widetilde{\mathcal{C}}$ is obtained from the first four columns of $\mathcal{C}$, which are linearly independent, by reordering:

$$\widetilde{\mathcal{C}} = \left[b_1|Ab_1|b_2|Ab_2\right] = \begin{bmatrix} 1 & 3 & 3 & 5 \\ 0 & -7 & 2 & 9 \\ 2 & 5 & -4 & -11 \\ -1 & 1 & 1 & 5 \end{bmatrix}.$$

The construction of $\widetilde{\mathcal{C}}$ also defines the numbers ν_1 and ν_2 as $\nu_1 = 2 = \nu_2$. We invert $\widetilde{\mathcal{C}}$ to obtain:

$$\widetilde{\mathcal{C}}^{-1} = \frac{1}{144}\begin{bmatrix} 36 & 45 & 45 & -18 \\ 8 & -8 & 8 & 24 \\ 28 & -37 & -53 & -78 \\ 0 & 18 & 18 & 36 \end{bmatrix}.$$

The vectors c_1 and c_2 and finally obtained as:

$$c_1^T = \frac{1}{144}\begin{bmatrix} 8 & -8 & 8 & 24 \end{bmatrix} \qquad c_2^T = \frac{1}{144}\begin{bmatrix} 0 & 18 & 18 & 36 \end{bmatrix}.$$

◁

The next example illustrates how symbolic abstractions of control systems can be used to synthesize controllers enforcing requirements stemming from infinite-state and finite-state specifications. It also shows the refinement process from the controller synthesized for the finite-state abstraction to the refined controller for the original infinite-state system.

Example 8.13. This is the longest example in the book. For the sake of readability it is divided in several sections.

Problem description

We consider an infra-red camera onboard of a mobile rover exploring the cold landscape of Mars. As the temperature on the Mars surface is typically below zero degrees Celsius, the rover is equipped with a heater that maintains the camera and other sensors warm. The objective is to synthesize the control software responsible for using the camera to take pictures. The requirements for this software module are:

1. the heater must be turned off when pictures are taken since the heater radiation masks the other infra-red sources;
2. the heater can only be turned off once per picture and must be turned on immediately after taking a picture to prevent the camera from freezing;
3. the camera must be controlled from the home position to the position required to take the picture and back to the home position in no more than 5 units of time.

Heater model

The heater is a shared resource since there are other software tasks that need to switch the heater on and off. This means that a request to turn the heater off may not be immediately granted. However, we assume that such request is granted after some time. To simplify the discussion, we assume this time to be one time step so that two consecutive requests to turn the heater off always result in the heater being turned off. Hence, we model the heater by the system S_{heat} in Figure 8.3. We can see that a request to turn the heater off, modeled by the input `off`, at the state $x_{heat\,1}$ may result in the heater remaining on, if the system moves to state $x_{heat\,2}$, or being turned off, if the system moves to state $x_{heat\,3}$. However, at the state $x_{heat\,2}$ a request to turn the heater off is immediately honored. The system in Figure 8.3 is a metaphor for other shared software/hardware resources which typically require more complex interaction protocols and are the source of several software bugs.

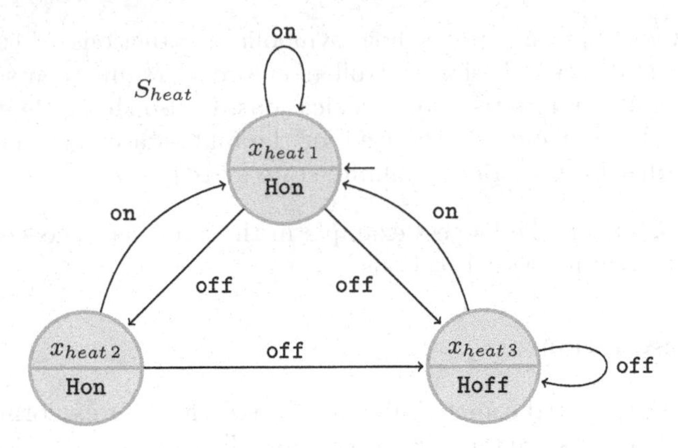

Fig. 8.3. Finite-state model S_{heat} for the heater.

Camera model

The software module to be designed, not only needs to interact correctly with the heater system, but it also needs to send continuous-time input signals to the camera pointing system. The later is described by a simple mechanical system including a friction term proportional to the angular velocity and an input describing the torque applied by a motor:

$$\dot{\theta} = \omega \tag{8.19}$$

$$\dot{\omega} = -\gamma\omega + u. \tag{8.20}$$

In this model, θ describes the angle that the camera makes with the rest position, that we call home position; ω represents the angular velocity of the camera; and u is the input. To keep our discussion simple, we let[2] $\gamma = 0$. Assuming that the input is updated every unit[3] of time and held constant between updates, we can explicitly integrate the differential equations (8.19) and (8.20) to obtain the discrete-time control system Σ defined by:

$$\theta(k+1) = \theta(k) + \omega(k) + \frac{1}{2}u(k)$$
$$\omega(k+1) = \omega(k) + u$$

with $k \in \mathbb{N}_0$ and where $\theta(t) = \theta(k)$ and $\omega(t) = \omega(k)$ at $t = k$, and $u(t) = u(k)$, for all $t \in [k, k+1[$. In order to apply Theorem 8.10, we model Σ as a system

[2] It is also possible to carry out the subsequent constructions with a nonzero friction term at the expense of a lengthier presentation.

[3] The choice of unit sampling period was made for simplicity of presentation only. Given any other sampling period, and assuming the inputs to be constant in between updates, we can always construct a discrete-time linear control system describing the evolution of a continuous-time linear control system at the sampling instants.

$S_Q(\Sigma)$ with Q an equivalence relation defined by a finite partition $\mathcal{P}$ adapted to Σ and adequate to express the specification. One possible choice for $\mathcal{P}$ is obtained by first constructing the vector c_1 from the requirement $c_1^T b_1 = 0$ as:

$$c_1^T = \left[1\ -\tfrac{1}{2}\right].$$

Using c_1^T and $c_1^T A$ we consider the partition $\mathcal{P}$ consisting of the following sets:

$$\begin{aligned}
P_1 &= \{(\theta,\omega) \in \mathbb{R}^2 \mid c_1^T x \le \varepsilon \wedge c_1^T x \ge -\varepsilon \wedge c_1^T Ax \le \varepsilon \wedge c_1^T Ax \ge -\varepsilon\} \\
P_2 &= \{(\theta,\omega) \in \mathbb{R}^2 \mid c_1^T x \le \varepsilon + \theta_{des} \wedge c_1^T x \ge -\varepsilon + \theta_{des} \wedge c_1^T Ax \le \varepsilon + \theta_{des} \\
&\qquad \wedge c_1^T Ax \ge -\varepsilon + \theta_{des}\} \\
P_3 &= \mathbb{R}^2 \backslash (P_1 \cup P_2)
\end{aligned}$$

where ε is a sufficiently small design parameter and θ_{des} is the angle to which the camera should be controlled to, in order to take the picture. The set P_1 can be made arbitrarily small by choosing ε sufficiently small and thus describes a small neighborhood of $x = (\theta,\omega) = (0,0)$. Similarly, the set P_2 describes a small neighborhood of $x = (\theta,\omega) = (\theta_{des},0)$. The sets P_1 and P_2 are depicted in Figure 8.4 for $\varepsilon = 0.5$ and $\theta_{des} = 4$. In terms of these sets, the objective is to control the camera from the home position P_1 to the orientation required for the picture and described by the set P_2. Once the picture has been taken, the camera should return to the home position P_1. Applying Algorithm 8.1 to this partition we obtain the new partition represented in Figure 8.4, for $\varepsilon = 0.5$ and $\theta_{des} = 4$, and defined by:

$$\begin{aligned}
P_1 &= \{(\theta,\omega) \in \mathbb{R}^2 \mid c_1^T x \le \varepsilon \wedge c_1^T x \ge -\varepsilon \wedge c_1^T Ax \le \varepsilon \wedge c_1^T Ax \ge -\varepsilon\} \\
P_2 &= \{(\theta,\omega) \in \mathbb{R}^2 \mid c_1^T x \le \varepsilon + \theta_{des} \wedge c_1^T x \ge -\varepsilon + \theta_{des} \wedge c_1^T Ax \le \varepsilon + \theta_{des} \\
&\qquad \wedge c_1^T Ax \ge -\varepsilon + \theta_{des}\} \\
P_3 &= \{(\theta,\omega) \in \mathbb{R}^2 \mid c_1^T Ax \le \varepsilon \wedge c_1^T Ax \ge -\varepsilon \wedge (c_1^T x > \varepsilon \vee c_1^T x < -\varepsilon)\} \\
P_4 &= \{(\theta,\omega) \in \mathbb{R}^2 \mid c_1^T Ax \le \varepsilon + \theta_{des} \wedge c_1^T Ax \ge -\varepsilon + \theta_{des} \wedge (c_1^T x > \varepsilon + \theta_{des} \\
&\qquad \vee c_1^T x < -\varepsilon + \theta_{des})\} \\
P_5 &= \mathbb{R}^2 \backslash (P_1 \cup P_2 \cup P_3 \cup P_4).
\end{aligned}$$

Figure 8.5 depicts the finite-state system S_{abs} bisimilar to $S_Q(\Sigma)$. System S_{abs} is obtained from the quotient of $S_Q(\Sigma)$ by the equivalence relation defined by the partition $\{P_1, P_2, P_3, P_4, P_5\}$ as follows. The set on inputs $U_{/R}$ is redefined to be $X_{/R}$ and every transition is relabeled with the destination state as the corresponding input. As discussed in Chapter 4, these changes result in a finite-state system S_{abs} that is still bisimilar to $S_Q(\Sigma)$. The states $x_{abs\, i}$ of S_{abs} correspond to the sets P_i for $i = 1, \ldots, 5$. This correspondence defines the bisimulation relation ${}_{abs}R_Q$ between S_{abs} and $S_Q(\Sigma)$. We note that since $S_Q(\Sigma)$ is deterministic, ${}_{abs}R_Q$ is also an alternating bisimulation relation.

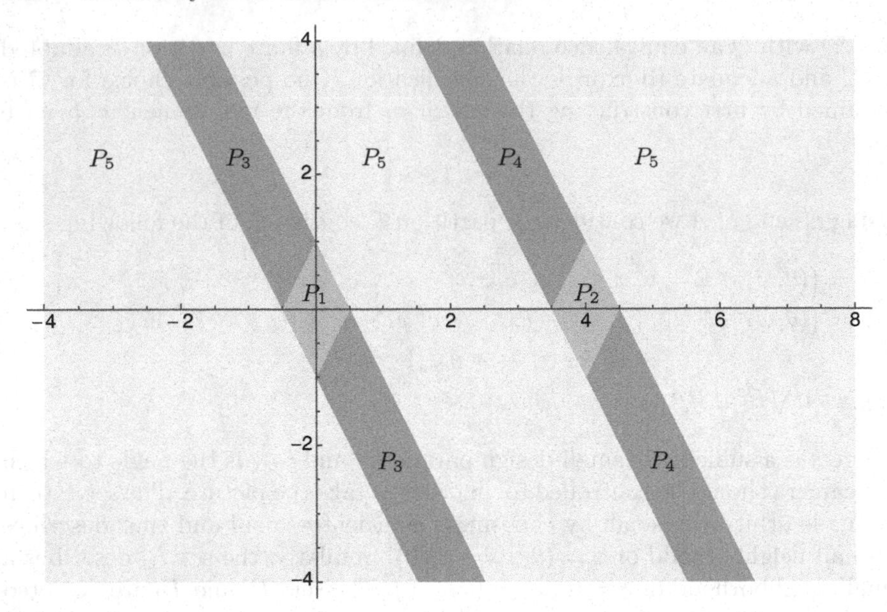

Fig. 8.4. Finite partition defining the finite-state system S_{abs} bisimilar to $S_Q(\Sigma)$ and depicted in Figure 8.5.

The composed system $S_{abs} \times S_{heat}$ describes the concurrent[4] operation of the camera pointing system and the heater system. For later use, we note that the alternating bisimulation relation ${}_{abs}R_Q$ can be extended to an alternating bisimulation relation ${}_{abs\,heat}R_{Q\,heat}$ between $S_{abs} \times S_{heat}$ and $S_Q(\Sigma) \times S_{heat}$ defined by $((x_{abs}, x_{heat}), (x, x'_{heat})) \in {}_{abs\,heat}R_{Q\,heat}$ if $(x_{abs}, x) \in {}_{abs}R_Q$ and $x_{heat} = x'_{heat}$.

Specification model

The software module to be designed can be regarded as a controller S_{cont} enforcing the previously described specification on $S_{abs} \times S_{heat}$. Noting that the infinite external behavior of $S_{abs} \times S_{heat}$ is a subset of $(Y_{abs} \times Y_{heat})^\omega$, we can describe the specification as the subset of $(Y_{abs} \times Y_{heat})^\omega$ defined by the strings:

$$(\texttt{home}, \texttt{Hon})(\texttt{pointed}, \texttt{Hoff})(\texttt{home}, \texttt{Hon})^\omega, \tag{8.21}$$

$$(\texttt{home}, \texttt{Hon})(\texttt{pointed}, \texttt{Hoff})(?, \texttt{Hon})(\texttt{home}, \texttt{Hon})^\omega, \tag{8.22}$$

$$(\texttt{home}, \texttt{Hon})(\texttt{pointed}, \texttt{Hoff})(?, \texttt{Hon})(?, \texttt{Hon})(\texttt{home}, \texttt{Hon})^\omega, \tag{8.23}$$

$$(\texttt{home}, \texttt{Hon})(\texttt{pointed}, \texttt{Hoff})(?, \texttt{Hon})(?, \texttt{Hon})(?, \texttt{Hon})(\texttt{home}, \texttt{Hon})^\omega, \tag{8.24}$$

[4] Recall from Chapter 1 that $S_{abs} \times S_{heat}$ denotes composition with respect to the trivial interconnection relation $X_{abs} \times X_{heat} \times U_{abs} \times U_{heat}$.

$$\texttt{(home,Hon)(?,Hon)(pointed,Hoff)(home,Hon)}^{\omega}, \tag{8.25}$$
$$\texttt{(home,Hon)(?,Hon)(pointed,Hoff)(?,Hon)(home,Hon)}^{\omega}, \tag{8.26}$$
$$\texttt{(home,Hon)(?,Hon)(pointed,Hoff)(?,Hon)(?,Hon)(home,Hon)}^{\omega}, \tag{8.27}$$
$$\texttt{(home,Hon)(?,Hon)(?,Hon)(pointed,Hoff)(home,Hon)}^{\omega}, \tag{8.28}$$
$$\texttt{(home,Hon)(?,Hon)(?,Hon)(pointed,Hoff)(?,Hon)(home,Hon)}^{\omega}, \tag{8.29}$$
$$\texttt{(home,Hon)(?,Hon)(?,Hon)(?,Hon)(pointed,Hoff)(home,Hon)}^{\omega}. \tag{8.30}$$

In the description of the specification, we used $(\texttt{home},\texttt{Hon})^{\omega}$ to describe the infinite repetition of the symbol (home, Hon) and we used (?, Hon) to denote any element of $Y_{abs} \times Y_{heat}$ with the second component equal to Hon. Note that all these strings start and end at the home position with the heater on; return to the home position in no more than 5 time units; visit the output

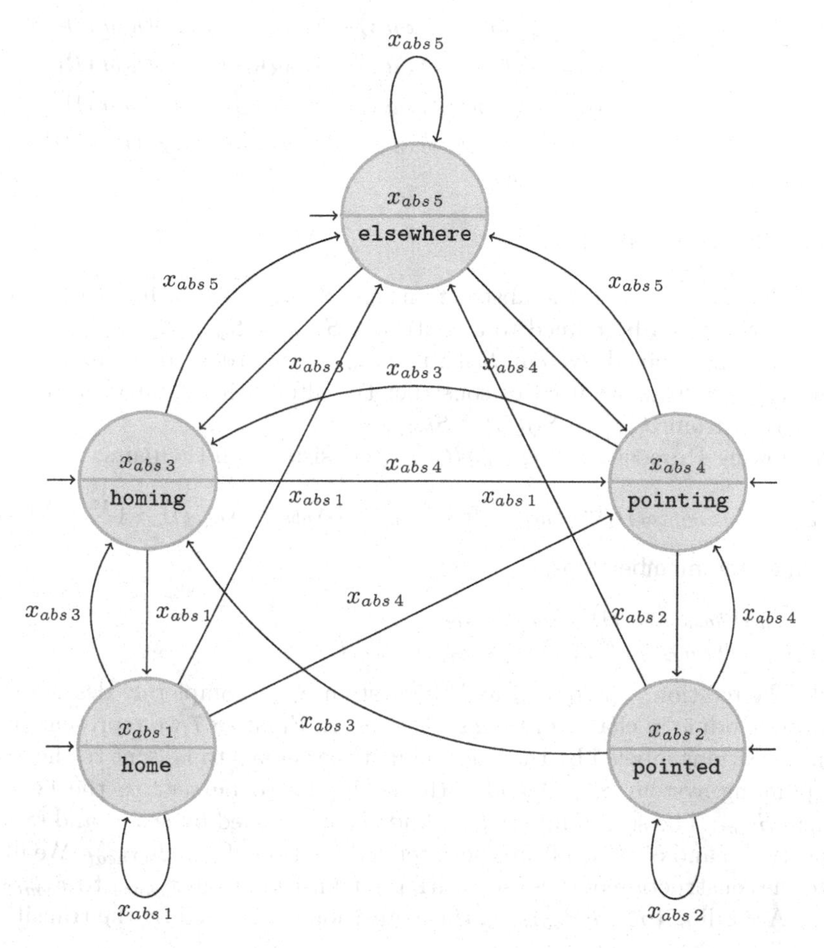

Fig. 8.5. Finite-state model S_{abs}, bisimilar to $S_Q(\Sigma)$, for the camera dynamics.

pointed with the heater off; and switch off the heater only once. Therefore, they describe all the behaviors of $S_{abs} \times S_{heat}$ that satisfy the specification.

Finite-state controller

The controller S_{cont} can now be synthesized by constructing an output deterministic finite-state system S_{spec} whose external behavior consists of the subset of $(Y_{abs} \times Y_{heat})^{\omega}$ defined by strings (8.21) through (8.30), and by solving a simulation game for system $S_{abs} \times S_{heat}$ and specification S_{spec}. Using the methods described in Chapter 6 we obtain the controller S_{cont}, represented in Figure 8.6 and the following alternating simulation relation from S_{cont} to $S_{abs} \times S_{heat}$:

$$\begin{aligned} {}_{cont}R_{abs\,heat} = \{ & (x_{cont\,1}, (x_{abs\,5}, x_{heat\,1})), (x_{cont\,2}, (x_{abs\,4}, x_{heat\,1})), \\ & (x_{cont\,3}, (x_{abs\,2}, x_{heat\,3})), (x_{cont\,4}, (x_{abs\,2}, x_{heat\,1})), \\ & (x_{cont\,5}, (x_{abs\,4}, x_{heat\,1})), (x_{cont\,6}, (x_{abs\,5}, x_{heat\,1})), \\ & (x_{cont\,7}, (x_{abs\,3}, x_{heat\,1})), (x_{cont\,8}, (x_{abs\,1}, x_{heat\,1})) \\ & (x_{cont\,9}, (x_{abs\,2}, x_{heat\,2})), (x_{cont\,10}, (x_{abs\,2}, x_{heat\,3}))\}. \end{aligned}$$

Controller refinement

Controller S_{cont} acts on the abstraction $S_{abs} \times S_{heat}$. According to Proposition 8.7, S_{cont} can be refined to a controller $S'_{cont} = S_{cont} \times_{\mathcal{F}} (S_{abs} \times S_{heat})$ acting on the refined system $S_Q(\Sigma) \times S_{heat}$. In order to compose S'_{cont} with $S_Q(\Sigma) \times S_{heat}$ we need to construct the alternating simulation relation ${}_{cont'}R_{Q\,heat}$ from S'_{cont} to $S_Q(\Sigma) \times S_{heat}$.

Following Proposition 8.7, ${}_{cont'}R_{Q\,heat}$ consists of all the pairs:

$$((x'_{cont}, (x_{abs}, x_{heat})), (x, x_{heat})) \in (X'_{cont} \times (X_{abs} \times X_{heat})) \times (X \times X_{heat})$$

for which the memberships:

1. $(x'_{cont}, (x_{abs}, x_{heat})) \in {}_{cont}R_{abs\,heat}$;
2. $((x_{abs}, x_{heat}), (x, x'_{heat})) \in {}_{abs\,heat}R_{Q\,heat}$,

hold. The relation ${}_{cont'}R_{Q\,heat}$ and the system S'_{cont} completely describe the software module in charge of taking pictures. In Figure 8.7 we represent S'_{cont} with transitions labeled by the continuous-time signals to be sent to the camera pointing system $S_Q(\Sigma)$ and with the inputs to be sent to the heating system S_{heat}. The sets of inputs U_{des} and U_0 are defined by (8.33) and (8.34), respectively, and obtained from the extended relation of ${}_{cont'}R_{Q\,heat}$. We illustrate the construction of these sets with the transition from $x_{cont\,1}$ to $x_{cont\,2}$ in S'_{cont}. According to ${}_{cont'}R_{abs\,heat}$, this transition is matched by the transition:

$$(x_{abs\,5}, x_{heat\,1}) \xrightarrow[abs\,heat]{x_{abs\,4},\mathsf{on}} (x_{abs\,4}, x_{heat\,1})$$

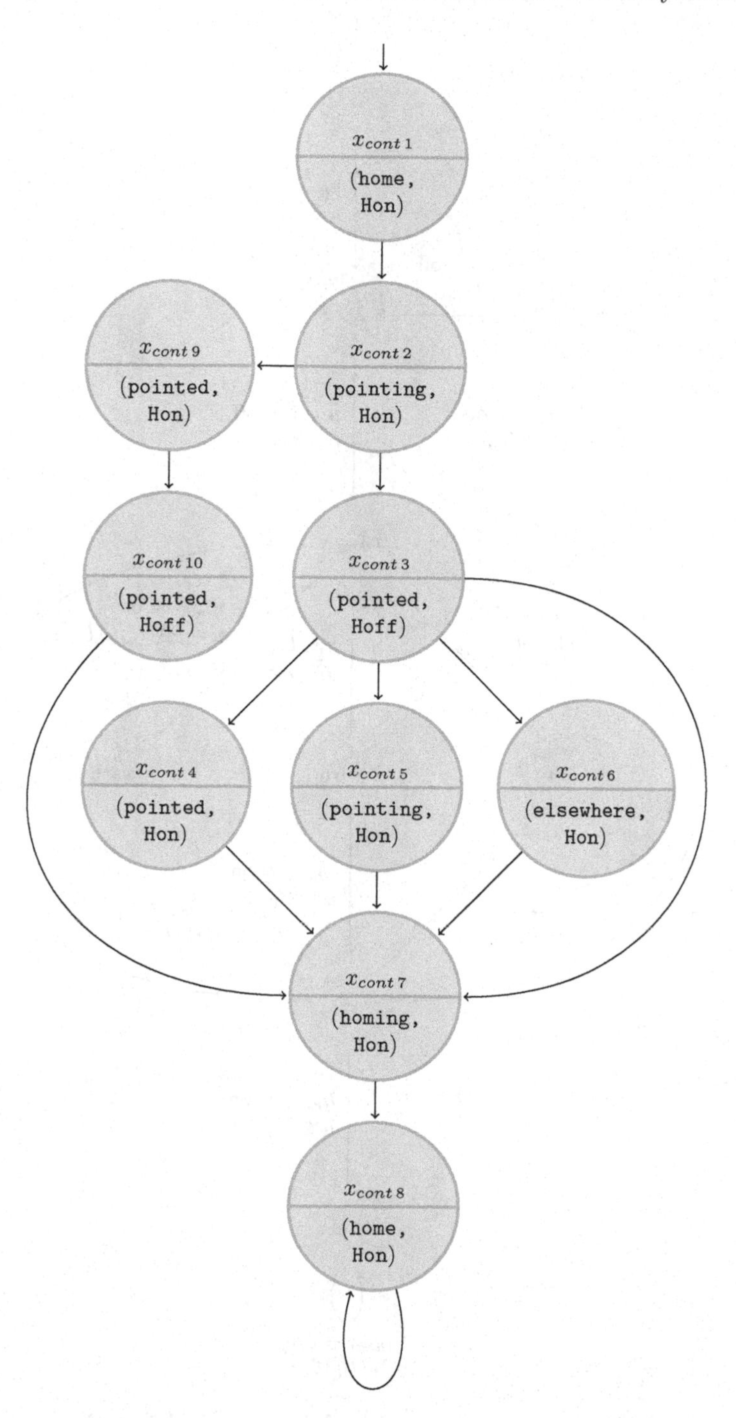

Fig. 8.6. Finite-state controller S_{cont} obtained as the solution of a simulation game for system $S_{abs} \times S_{heat}$ and specification system S_{spec}.

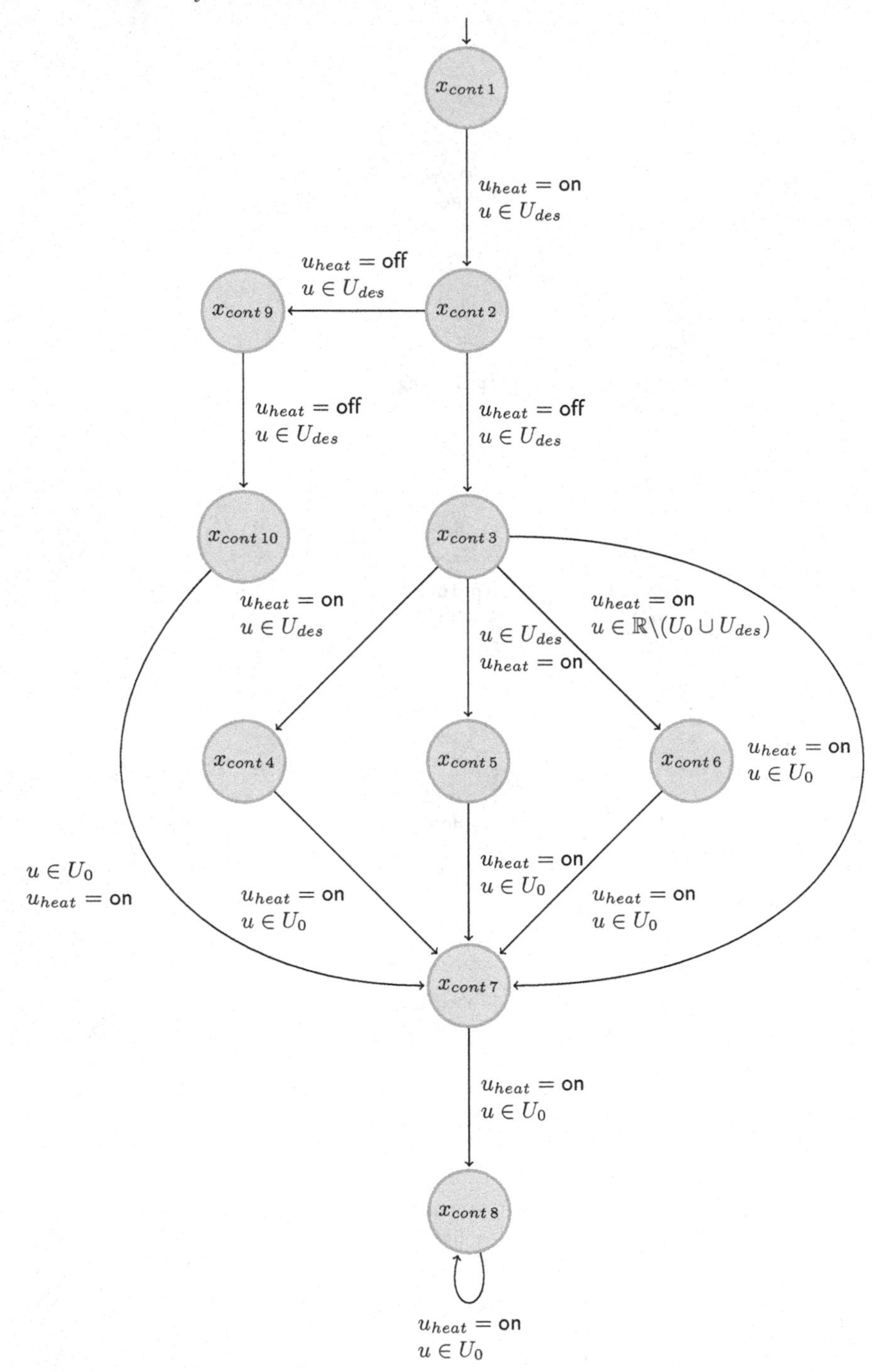

Fig. 8.7. Controller S'_{cont} with transitions labeled with the continuous-time and discrete-time inputs sent to $S_Q(\Sigma) \times S_{heat}$.

in $S_{abs} \times S_{heat}$ and according to ${}_{abs\,heat}R_{Q\,heat}$, it is also matched by all the transitions $x \xrightarrow{u} x'$ in $S_Q(\Sigma)$ for which $x \in P_5$ and $x' = Ax + Bu \in P_4$. Using the definition of the set P_4, the inclusion $Ax+Bu \in P_4$ can be expressed as the set of points $x \in P_5$ for which there exists $u \in \mathbb{R}$ satisfying:

$$c_1^T A^2 x + c_1^T ABu \leq \varepsilon + \theta_{des} \wedge c_1^T A^2 x + c_1^T ABu \geq -\varepsilon + \theta_{des} \quad (8.31)$$

$$c_1^T Ax + c_1^T Bu > \varepsilon + \theta_{des} \vee c_1^T Ax + c_1^T Bu < -\varepsilon + \theta_{des} \quad (8.32)$$

Noting that $c_1^T B = 0$, (8.32) simplifies to $c_1^T Ax > \varepsilon+\theta_{des} \vee c_1^T Ax < -\varepsilon+\theta_{des}$. Since all the points in P_5 already satisfy these inequalities we analyze (8.31). Using $c_1^T AB = 1$ we obtain the explicit description:

$$u \in U_{des}(\theta, \omega) = \left[-\theta - \frac{3}{2}\omega + \theta_{des} - \varepsilon, -\theta - \frac{3}{2}\omega + \theta_{des} + \varepsilon\right] \quad (8.33)$$

for the set of inputs labeling the transition $x \xrightarrow{u} x'$ in $S_Q(\Sigma)$. Similar computations lead to the set of inputs:

$$u \in U_0(\theta, \omega) = \left[-\theta - \frac{3}{2}\omega - \varepsilon, -\theta - \frac{3}{2}\omega + \varepsilon\right] \quad (8.34)$$

labeling transitions in $S_Q(\Sigma)$ matching the transition from $x_{cont\,5}$ to $x_{cont\,7}$ in S'_{cont}. Using the sets U_{des} and U_0 we can describe all the inputs that need to be sent to the camera pointing system, as depicted in Figure 8.7.

Controller S'_{cont} uses state measurements from $S_Q(\Sigma) \times S_{heat}$ in two different ways: in order to determine which inputs are to be sent to $S_Q(\Sigma)$, since the sets U_{des} and U_0 depend on the current infinite state; and in order to determine which state of S_{heat} is reached when the input off is applied at the state $x_{heat\,1}$ of S_{heat}, corresponding to state $x_{cont\,2}$ in S'_{cont}. The system represented in Figure 8.7 can also be seen as a hybrid controller since it requires infinite-state and finite-state measurements to operate and it generates continuous-time and discrete-time output signals to control $S_Q(\Sigma) \times S_{heat}$. $\triangleleft$

8.4 Continuous-time multi-affine control systems

In this section we describe a different abstraction technique applicable to a large class of control systems in continuous-time.

Before recalling the notion of affine function, we introduce some notation that is only used in this section. Given a map $f : A_1 \times A_2 \times A_3 \to B$, we denote by $f_{\widehat{a_2}} : A_2 \to B$ the map $f(a_1, \cdot, a_3) : A_2 \to B$ obtained from f by fixing the entries a_1 and a_3. Similarly, $f_{\widehat{a_1}} : A_1 \to B$ and $f_{\widehat{a_3}} : A_3 \to B$ denote the maps obtained from f by fixing a_2 and a_3 in the first case and a_1 and a_2 in the second case. Given a point $a \in A$, the membership function for a is denoted by $\chi_a : A \to \{0, 1\}$ and satisfies $\chi_a(x) = 1$ if $x = a$ and $\chi_a(x) = 0$ if $x \neq a$.

Affine functions are characterized by the following property.

Definition 8.14 (Affine function). *A map $f : \mathbb{R} \to \mathbb{R}$ is said to be* affine *when for every $x, y \in \mathbb{R}$ and for every $\alpha, \beta \in \mathbb{R}$ satisfying $\alpha + \beta = 1$ the following equality holds:*

$$f(\alpha x + \beta y) = \alpha f(x) + \beta f(y).$$

The number $\alpha x + \beta y$ is said to be an affine combination of x and y when $\alpha + \beta = 1$. The constraint $\alpha + \beta = 1$ can be expressed using a single variable $\lambda \in \mathbb{R}$ to define the affine combination of x and y as $(1 - \lambda)x + \lambda y$. A set $Z \subseteq \mathbb{R}^n$ is said to be convex when it contains all the affine combinations of points $x, y \in Z$ with $\lambda > 0$. Consider now two intervals $E_1 =]a_1, b_1[\subset \mathbb{R}$ and $E_2 =]a_2, b_2[\subset \mathbb{R}$. A point $x_1 \in E_1$ can be written as the affine combination $(1 - \lambda_1)a_1 + \lambda_1 b_1$ of the vertices a_1 and b_1 of E_1 where λ_1 is obtained by solving $x_1 = (1 - \lambda_1)a_1 + \lambda_1 b_1$ to obtain:

$$\lambda_1 = \frac{x_1 - a_1}{b_1 - a_1}. \tag{8.35}$$

Similarly, a point $x_2 \in E_2$ can always be written as the affine combination $x_2 = (1 - \lambda_2)a_2 + \lambda_2 b_2$ with:

$$\lambda_2 = \frac{x_2 - a_2}{b_2 - a_2}. \tag{8.36}$$

This idea can be generalized to points $x \in E_1 \times E_2 \subset \mathbb{R}^2$. In this case we can still write x as an affine combination of the vertices:

$$v_{aa} = \begin{bmatrix} a_1 \\ a_2 \end{bmatrix}, \quad v_{ab} = \begin{bmatrix} a_1 \\ b_2 \end{bmatrix}, v_{ba} = \begin{bmatrix} b_1 \\ a_2 \end{bmatrix}, \quad v_{bb} = \begin{bmatrix} b_1 \\ b_2 \end{bmatrix}$$

of the rectangle $E_1 \times E_2$:

$$x = (1-\lambda_1)(1-\lambda_2)v_{aa} + (1-\lambda_1)(\lambda_2)v_{ab} + (\lambda_1)(1-\lambda_2)v_{ba} + (\lambda_1)(\lambda_2)v_{bb}. \tag{8.37}$$

Although it is not difficult to see that the preceding coefficients do add to one, it may not be immediately obvious that λ_1 and λ_2 are still given by (8.35) and (8.36), respectively. This can be verified by simply substituting (8.35) and (8.36) in (8.37). The affine combination in (8.37) is not unique since a point in $E_1 \times E_2$ can be written as an affine combination of three vertices only. However, the particular affine combination in (8.37) enables us to consider non-affine functions while still retaining some of the properties of affine functions. Consider a function $f : \mathbb{R}^2 \to \mathbb{R}$ such that $f_{\widehat{x_1}} : \mathbb{R} \to \mathbb{R}$ and $f_{\widehat{x_2}} : \mathbb{R} \to \mathbb{R}$ are affine maps. Then, $f(x)$ for $x \in E_1 \times E_2$ can be obtained as the affine combination of $f(v_{aa})$, $f(v_{ab})$, $f(v_{ba})$ and $f(v_{bb})$. To see this, introduce first the auxiliary variables:

$$z = (1 - \lambda_2)v_{aa} + (\lambda_2)v_{ab} \qquad w = (1 - \lambda_2)v_{ba} + (\lambda_2)v_{bb}.$$

Using z and w we can write (8.37) as:

$$x = (1-\lambda_1)z + (\lambda_1)w$$

and:

$$\begin{aligned} f(x) &= f\big((1-\lambda_1)z + (\lambda_1)w\big) \\ &= f_{\widehat{x_1}}\big((1-\lambda_1)z + (\lambda_1)w\big) \\ &= (1-\lambda_1)f_{\widehat{x_1}}(z) + (\lambda_1)f_{\widehat{x_1}}(w) \\ &= (1-\lambda_1)f(z) + (\lambda_1)f(w). \end{aligned} \tag{8.38}$$

Note now that:

$$\begin{aligned} f(z) &= f\big((1-\lambda_2)v_{aa} + (\lambda_2)v_{ab}\big) \\ &= f_{\widehat{x_2}}\big((1-\lambda_2)v_{aa} + (\lambda_2)v_{ab}\big) \\ &= (1-\lambda_2)f_{\widehat{x_2}}(v_{aa}) + (\lambda_2)f_{\widehat{x_2}}(v_{ab}) \\ &= (1-\lambda_2)f(v_{aa}) + (\lambda_2)f(v_{ab}). \end{aligned} \tag{8.39}$$

Similarly:

$$\begin{aligned} f(w) &= f\big((1-\lambda_2)v_{ba} + (\lambda_2)v_{bb}\big) \\ &= (1-\lambda_2)f(v_{ba}) + (\lambda_2)f(v_{bb}). \end{aligned} \tag{8.40}$$

Finally, we combine (8.38), (8.39) and (8.40) to obtain:

$$\begin{aligned} f(x) = (1-\lambda_1)(1-\lambda_2)f(v_{aa}) + (1-\lambda_1)(\lambda_2)f(v_{ab}) \\ +(\lambda_1)(1-\lambda_2)f(v_{ba}) + (\lambda_1)(\lambda_2)f(v_{bb}). \end{aligned} \tag{8.41}$$

We now generalize this simple idea beyond rectangles in $\mathbb{R}^2$.

Definition 8.15 (n-rectangle). *A* n-rectangle E *in* $\mathbb{R}^n$ *is a set defined by:*

$$E = \prod_{i=1}^{n}]a_i, b_i[$$

where $a_i, b_i \in \mathbb{R}$ *satisfy* $a_i < b_i$ *for* $i = 1, \ldots, n$. *The set of vertices of an* n*-rectangle is denoted by* $\mathcal{V}(E)$ *and defined by:*

$$\mathcal{V}(E) = \big\{x \in \mathbb{R}^n \mid x_i \in \{a_i, b_i\}\big\}.$$

Definition 8.16 (Multi-affine function). *A map* $f : \mathbb{R}^n \to \mathbb{R}$ *is said to be* multi-affine *when for each* x_i, $i = 1, \ldots, n$, $f_{\widehat{x_i}} : \mathbb{R} \to \mathbb{R}$ *is affine. A map* $f : \mathbb{R}^n \to \mathbb{R}^m$ *is multi-affine, when for each* $i = 1, \ldots, m$ *the map* $f_i : \mathbb{R}^n \to \mathbb{R}$ *is multi-affine.*

The following result generalizes (8.41) to arbitrary dimensions.

Proposition 8.17. *Let E be an n-rectangle in $\mathbb{R}^n$ and $f : \mathbb{R}^n \to \mathbb{R}^m$ a multi-affine function. The following holds:*

$$x \in E \quad \Longrightarrow \quad f(x) = \sum_{v \in \mathcal{V}(E)} \lambda_v f(v), \quad \sum_{v \in \mathcal{V}(E)} \lambda_v = 1.$$

Multi-affine maps are completely determined by the values they assume on the vertices of n-rectangles in $\mathbb{R}^n$.

Proposition 8.18. *Let E be an n-rectangle in $\mathbb{R}^n$ and consider a function $g : \mathcal{V}(E) \to \mathbb{R}^m$. There exists one and only one multi-affine function $f : \mathbb{R}^n \to \mathbb{R}^m$ such that $f|_{\mathcal{V}(E)} = g$. Moreover, f is given by:*

$$f(x) = \sum_{v \in \mathcal{V}(E)} \prod_{i=1}^{n} \left(\frac{x_i - a_i}{b_i - a_i} \right)^{\chi_{b_i}(v_i)} \left(\frac{b_i - x_i}{b_i - a_i} \right)^{\chi_{a_i}(v_i)} g(v). \tag{8.42}$$

The function f in (8.42) was obtained as a straightforward generalization of equality (8.41) thus guaranteeing that f is multi-affine and that $f|_{\mathcal{V}(E)} = g$. Moreover, f is unique since if we assume the existence of another multi-affine function $f' : \mathbb{R}^n \to \mathbb{R}^m$ satisfying $f'|_{\mathcal{V}(E)} = g$ we conclude that $f - f'$ is still multi-affine and satisfies $(f - f')|_{\mathcal{V}(E)} = g - g = 0$. It then follows from Proposition 8.17 that $f - f'$ is the function identically zero and thus $f = f'$.

Multi-affine control systems are characterized by restrictions on U and f.

Definition 8.19 (Multi-affine control system). *A control system $(\mathbb{R}^n, \mathcal{U}, f)$ is said to be* multi-affine *if f is of the form $f(x,u) = g(x) + Bu$ with $B \in \mathbb{R}^{n \times m}$ and $g : \mathbb{R}^n \to \mathbb{R}^n$ a multi-affine function.*

Multi-affine control systems are also denoted by the more informative quadruple $\Sigma = (\mathbb{R}^n, \mathcal{U}, g, B)$. The definition of multi-affine control systems requires B to be constant. When B is not constant but a multi-affine function of x, the results in this section are still applicable by considering the extended system:

$$\begin{bmatrix} \dot{x} \\ \dot{u} \end{bmatrix} = \begin{bmatrix} g(x) + B(x)u \\ 0 \end{bmatrix} + \begin{bmatrix} 0 \\ I_m \end{bmatrix} v \tag{8.43}$$

where the state is now $(x,u) \in \mathbb{R}^n \times \mathbb{R}^m$, the input is $v \in \mathbb{R}^m$, and I_m is the $m \times m$ identity matrix.

Whenever a control system is multi-affine, we can exploit its multi-affine properties to obtain symbolic abstractions based on partitions of state space induced by n-rectangles. The construction of such abstractions is based on a careful analysis of the interplay between the functions g and B, defining the dynamics, and the facets and vertices of n-rectangles. Recall that a facet of an n-rectangle E is the intersection of $\overline{E}$ with an affine subspace of dimension $n-1$ defined by $x_i = a_i$ or $x_i = b_i$. Given a vertex $v \in \mathcal{V}(E)$ we denote by

$\mathcal{F}(v)$ the set of all facets containing v. Moreover, to each facet F defined by $x_i = a_i$ we associate a normal vector η_F defined by $\eta_{Fj} = 0$ for $j \neq i$ and $\eta_{Fj} = -1$ for $j = i$. Similarly, to a facet defined by $x_i = b_i$ we associate a normal vector defined by $\eta_{Fj} = 0$ for $j \neq i$ and $\eta_{Fj} = 1$ for $j = i$. When F is a facet defined by $x_i = a_i$, F^o denotes the opposite facet defined by $x_i = b_i$. Similarly, if F denotes the facet $x_i = b_i$, F^o denotes the opposite facet defined by $x_i = a_i$. For each n-rectangle we consider the following two problems:

Problem 8.20 (Rectangular invariant). Let $\Sigma = (\mathbb{R}^n, \mathcal{U}, g, B)$ be a multi-affine control system and let E be an n-rectangle. The rectangular invariant problem consists in determining if there exists a multi-affine feedback control law $k : \mathbb{R}^n \to \mathbb{R}^m$ such that any solution ξ of the closed-loop dynamical system $(\mathbb{R}^n, g + Bk)$ satisfies:

$$\xi(0) \in E \implies \xi(t) \in E \quad \forall t \in \mathbb{R}^+.$$

Problem 8.21 (Control to facet). Let $\Sigma = (\mathbb{R}^n, \mathcal{U}, g, B)$ be a multi-affine control system, let E be an n-rectangle, and let F be a facet of E. The control to facet problem consists in determining if there exists a multi-affine feedback control law $k : \mathbb{R}^n \to \mathbb{R}^m$ such that for any solution ξ of the closed-loop dynamical system $(\mathbb{R}^n, g + Bk)$ for which $\xi(0) \in E$ there exists a time $\tau \in \mathbb{R}^+$ satisfying:

1. $\xi(t) \in E$ for $t \in [0, \tau[$;
2. $\xi(\tau) \in F$;
3. $\xi(t) \notin E \cup F$ for $t \in]\tau, \tau + \varepsilon]$ and some $\varepsilon \in \mathbb{R}^+$.

Once we know how to solve the preceding problems, the construction of finite-state abstractions is conceptually simple. We use n-rectangles as states and we place a transition from: a rectangle E to itself when Problem 8.20 has a solution for E; a rectangle E to rectangle E' when they share a facet F and Problem 8.21 has a solution for E and F.

The rectangular invariant problem admits the following solution.

Theorem 8.22. *Let Σ be a multi-affine control system and let E be a n-rectangle. The rectangular invariant problem admits a solution if the following sets are non-empty:*

$$U_E(v) = \bigcap_{F \in \mathcal{F}(v)} \{u \in \mathbb{R}^m \mid \eta_F^T(g(v) + Bu) < 0\} \tag{8.44}$$

for all $v \in \mathcal{V}(E)$.

Proof. We first show how to construct the multi-affine feedback from the sets U_E which are assumed to be nonempty. Let $h : \mathcal{V}(E) \to \mathbb{R}^m$ be a function satisfying $h(v) \in U_E(v)$. This function can be constructed by picking an element $u_v \in U_E(v)$ for every $v \in \mathcal{V}(E)$ and setting $h(v) = u_v$. By Proposition 8.18

there exists a unique multi-affine function $k : \mathbb{R}^n \to \mathbb{R}^m$ satisfying $k|_{\mathcal{V}(E)} = h$. Let us consider now the closed loop system defined by the differential equation $\dot{\xi} = f(\xi)$ with $f(x) = g(x) + Bk(x)$. The function f is multi-affine and since $k(v) \in U_E(v)$ we have $\eta_F^T f(v) < 0$ for every facet F of E. Consider now a point $x \in F$. By writing x as an affine combination of the vertices of F we conclude that $\eta_F^T f(v) < 0$ for every vertex v of F implies $\eta_F^T f(x) < 0$. It now follows from continuity of the function $\eta_F^T f$ that $\eta_F^T f(x) < 0$ for $x \in F$ implies $\eta_F^T f(y) < 0$ for every y in a neighborhood of x and thus in a neighborhood $\mathcal{N}$ of F. Consequently, no trajectory ξ of Σ starting inside E can touch F since this would require that for some τ, $\xi(\tau) \in \mathcal{N}$. However, $\frac{d}{dt}\eta_F^T \xi\big|_{t=\tau} = \eta_F^T f(\xi(\tau)) < 0$ which implies that $\eta_F^T f$ is decreasing and cannot reach F. Since this argument holds for all the facets of E, trajectories of S starting inside E cannot leave E. □

Theorem 8.22 provides us with very friendly conditions to test the existence of a solution to Problem 8.20. It suffices to check non-emptiness of the sets $U_E(v)$ which are described by affine inequalities. Moreover, when these sets are non-empty, a multi-affine feedback can be constructed by using expression (8.42) with $g(v)$ being any element of the set $U_E(v)$.

The control to facet problem also admits a simple solution.

Theorem 8.23. *Let Σ be a multi-affine control system, let E be an n-rectangle and let F be a facet of E. The control to facet problem admits a solution if the following sets are non-empty:*

$$U_E(v) = \bigcap_{G \in \mathcal{F}(v)} \{u \in \mathbb{R}^m \mid \eta_G^T(g(v) + Bu) < 0\}$$

for all $v \in \mathcal{V}(E)$ such that $F \notin \mathcal{F}(v)$, and:

$$U_E(v) = \bigcap_{G \in \mathcal{F}(v), G \neq F} \{u \in \mathbb{R}^m \mid \eta_G^T(g(v) + Bu) < 0 \ \wedge \ \eta_F^T(g(v) + Bu) > 0\}$$

for all $v \in \mathcal{V}(E)$ such that $F \in \mathcal{F}(v)$.

Proof. As in the proof of Theorem 8.22 we construct the feedback $k : \mathbb{R}^n \to \mathbb{R}^m$ as the unique multi-affine map satisfying $k|_{\mathcal{V}(E)} = h$ for some function $h : \mathcal{V}(E) \to \mathbb{R}^m$ with $h(v) \in U_E(v)$. Consider now any facet $G \neq F$. Since $\eta_G^T(g(v)+Bk(v)) < 0$ for every vertex v of F, it follows $\eta_G^T(g(x) + Bk(x)) < 0$ for every $x \in G$. As shown in the proof of Theorem 8.22, trajectories starting inside E cannot touch G. To conclude the proof we need to show that every trajectory starting inside E leaves E through F in finite time. To do this, we note that $\eta_F^T(g(v) + Bk(v)) > 0$ for every $v \in \mathcal{V}(E)$ implies, by writing $x \in \overline{E}$ as a linear combination of the vertices of E, that $\eta_F^T(g(x) + Bk(x)) > 0$ for every $x \in \overline{E}$. Let now $\delta = \min_{x \in \overline{E}} \eta_F^T(g(x)+Bk(x))$ which is well defined since $\overline{E}$ is a compact set and $\eta_F^T(g(x) + Bk(x))$ is a continuous function of $x \in \overline{E}$.

If we consider the evolution of $\eta_F^T \xi$ for a trajectory ξ such that $\xi(0) = x \in E$, it follows that:

$$\frac{d}{dt}\eta_F^T \xi = \eta_F^T \big(g(\xi) + Bk(\xi)\big) \geq \delta$$

for all t such that $\xi(t) \in E$. Therefore:

$$\eta_F^T \xi(t) \geq \eta_F^T \xi(0) + \delta t$$

for all t such that $\xi(t) \in E$. This means that is takes no longer than $(\eta_F^T y - \eta_F^T \xi(0))/\delta$ units of time for ξ to reach a point $y \in F$. Let us denote by τ the exact time at which $\xi(\tau) \in F$. It now follows from $\eta_F^T(g(y)+Bk(y)) > \delta$, for any $y \in F$, the existence of $\varepsilon > 0$ such that $\xi(t) \notin E \cup F$ for $t \in]\tau, \tau + \varepsilon]$. □

Example 8.24. Consider the Lotka-Volterra predator-prey model:

$$\dot{\xi}_1 = -\xi_1 + \xi_1 \xi_2 - v \tag{8.45}$$
$$\dot{\xi}_2 = \xi_2 - \xi_1 \xi_2 \tag{8.46}$$

where $\xi_1(t) \in \mathbb{R}_0^+$ describes the number of predators, $\xi_2(t) \in \mathbb{R}_0^+$ describes the number of preys, and $v(t) \in \mathbb{R}_0^+$ describes the effect[5] of pesticides repressing the predators at time $t \in \mathbb{R}$. We assume that the initial population lies in the rectangle $]5,6[\times]5,6[$. Without applying any control, the number of preys will diminish at the expense of increasing the number of predators. Assume now that we want to force the number of predators to remain in the interval $]5,6[$ while reducing the number to preys to a level below 2. This objective can be formulated as an instance of the control to facet problem. The objective is to synthesize a multi-affine control law ensuring that all the trajectories starting in the rectangle $E =]5,6[\times]2,6[$ exit through the facet $[5,6] \times \{2\}$. For this example we have four vertices:

$$\mathcal{V}(E) = \{(5,6), (6,6), (5,2), (6,2)\}.$$

The set $U_E(5,6)$ can be computed as:

$$\begin{aligned} U_E(5,6) &= \{u \in \mathbb{R}_0^+ \mid -5 + 5 \cdot 6 - u > 0 \wedge 6 - 5 \cdot 6 < 0\} \\ &= \{u \in \mathbb{R}_0^+ \mid u < 25\} \end{aligned}$$

and similar computations provide:

$$\begin{aligned} U_E(6,6) &= \{u \in \mathbb{R}_0^+ \mid u > 30\}, \\ U_E(5,2) &= \{u \in \mathbb{R}_0^+ \mid u < 5\}, \\ U_E(6,2) &= \{u \in \mathbb{R}_0^+ \mid u > 6\}. \end{aligned}$$

[5] A more accurate model is $\dot{\xi}_1 = -\xi_1 + \xi_1 \xi_2 - \xi_1 v$, describing the decrease of predators as being proportional to the product of the quantity of pesticides with the number of predators. This more general model can be converted into the multi-affine format by considering the extended system described in (8.43).

Since all the sets are non-empty, the problem is solvable. We choose the following points from each set:

$$15 \in U_E(5,6), \quad 40 \in U_E(6,6), \quad 0 \in U_E(5,2), \quad 15 \in U_E(6,2),$$

which define the desired multi-linear feedback $k : \mathbb{R}_0^+ \times \mathbb{R}_0^+ \to \mathbb{R}_0^+$ uniquely. Using expression (8.42) we obtain:

$$k(x_1, x_2) = -\frac{115}{2} + 10x_1 - \frac{35}{4}x_2 + \frac{5}{2}x_1x_2. \tag{8.47}$$

The resulting closed loop dynamical system is displayed in Figure 8.8 where it can be appreciated that $g(x) + Bk(x)$ always points inside E except at the exit facet. ◁

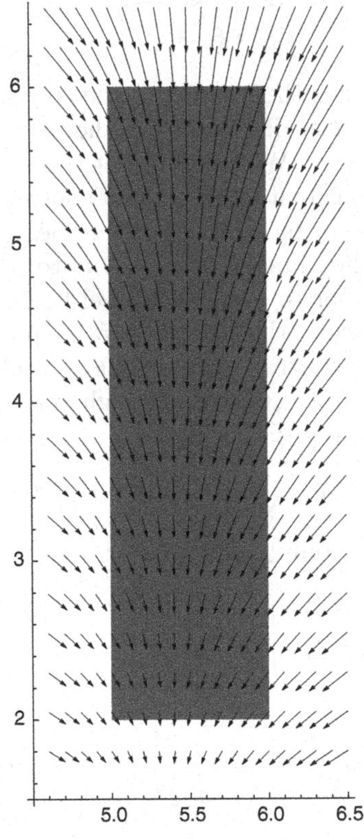

Fig. 8.8. Dynamical system resulting from applying the multi-affine feedback law (8.47) to the control system defined by (8.45) and (8.46). The rectangle E is dark-colored.

The solutions to Problems 8.20 and 8.21 given by Theorems 8.22 and 8.23 enable us to construct finite-state models for multi-affine systems. Although a collection of n-rectangles $\mathcal{E}$ does not form a partition of $\mathbb{R}^n$, under certain circumstances we can complete $\mathcal{E}$ to partition although not uniquely. We say that a partition $\mathcal{P}$ of $\mathbb{R}^n$ is a *completion* of a collection $\mathcal{E}$ of n-rectangles when:

1. $\forall E \in \mathcal{E}\ \exists P \in \mathcal{P}\quad \text{int}\, E = \text{int}\, P$;
2. $\forall P \in \mathcal{P}\ \exists E \in \mathcal{E}\quad \text{int}\, P = \text{int}\, E$.

Moreover, we denote by $P(E)$ the element of $\mathcal{P}$ for which $\text{int}P(E) = \text{int}E$.

Definition 8.25. *Consider a multi-affine control system Σ, let $\mathcal{P}$ be a partition of $\mathbb{R}^n$ that is a completion of a collection of n-rectangles $\mathcal{E}$, and let Q be the equivalence relation defined by $\mathcal{P}$. The finite-state system associated with Σ and $\mathcal{E}$, denoted by $S_{\mathcal{E}}(\Sigma) = (X_{\mathcal{E}}, U_{\mathcal{E}}, \underset{\mathcal{E}}{\longrightarrow}, Y_{\mathcal{E}}, H_{\mathcal{E}})$, consists of:*

- $X_{\mathcal{E}} = \mathcal{E}$;
- $U_{\mathcal{E}} = \mathcal{E}$;
- $x \overset{x'}{\underset{\mathcal{E}}{\longrightarrow}} x'$ *if any of the following two conditions holds:*
  1. *the equality $x = x'$ is satisfied and Problem 8.20 is solvable for Σ with n-rectangle x;*
  2. *the n-rectangles x and x' share a facet F and Problem 8.21 is solvable for Σ with n-rectangle x and facet F;*
- $Y_{\mathcal{E}} = \mathbb{R}^n/Q$;
- $H_{\mathcal{E}}(E) = P(E)$.

Definition 8.25, Theorem 8.22, and Theorem 8.23 have the following immediate consequence.

Theorem 8.26. *Consider a multi-affine control system Σ, let $\mathcal{P}$ be a partition of $\mathbb{R}^n$ that is a completion of a collection of n-rectangles $\mathcal{E}$, and let Q be the equivalence relation defined by $\mathcal{P}$. The relation $R \subseteq \mathcal{E} \times \mathbb{R}^n$ defined by:*

$$R = \{(E, x) \in \mathcal{E} \times \mathbb{R}^n \mid x \in E\}$$

is a simulation relation from $S_{\mathcal{E}}(\Sigma)$ to $S_Q(\Sigma)$.

The simulation relation from $S_{\mathcal{E}}(\Sigma)$ to $S_Q(\Sigma)$ is also an alternating simulation relation since both systems are deterministic. Therefore, the abstract system $S_{\mathcal{E}}(\Sigma)$ can be used to synthesize a controller that can be refined to a controller acting on $S_Q(\Sigma)$ as discussed in Section 8.2.

The abstraction technique based on the solutions to Problem 8.20 and Problem 8.21 is closely related to the sign based abstractions discussed in Chapter 7. Consider the solution to the rectangular invariant problem given in Theorem 8.22. Condition (8.44) can be understood as a sign condition of the form $\text{sign}(L_{g(x)+Bu}p) = -1$ for the function $p(x) = \eta_F^T x$. Due to the special geometry of the considered sets, n-rectangles, and due to the multi-affine nature of g, the analysis of the sign conditions only needs to be performed at the vertices of the n-rectangles. This fact makes this technique very appealing from a computational perspective. However, the construction of sign based abstractions can be generalized to control systems thus allowing to deal with more general differential equations and more general partitions.

8.5 Notes

Our treatment of controller refinement touched very superficially upon the relation between (alternating) simulation relations and composition. A more detailed treatment of these ideas can be found, *e.g.*, in [Mil89].

Adapted sets first appeared in the paper [TP03] discussing the possibility of model checking Linear Temporal Logic (LTL) formulas over controllable linear systems. A generalization of these results, allowing for the synthesis of controllers enforcing LTL specifications, is described in [TP06]. The results reported in the preceding two references require controllability. The discussion in Section 8.3 clearly separates the role of adapted sets from the role of controllability in the existence and construction of symbolic bisimilar models.

The proof of Theorem 8.4 can be found in any linear systems theory book such as [AM97].

The line of research on abstractions of multi-affine control systems and n-rectangles first started with piecewise linear systems defined on simplices and rectangles [HvS01]. The technique was then extended to multi-affine systems on n-rectangles [BH06]. The discussion in Section 8.4 treats rectangles as open sets while in [BH06] rectangles are treated as closed sets. The use of closed sets makes the conditions appearing in Theorem 8.23 not only sufficient but also necessary. However, treating rectangles as open sets allowed us to relate the abstraction to the original system through a simulation relation. Other results on this abstraction technique can be found in [HvS04, HCvS06, RB06, RB07].

The use of finite-state abstractions for control design has been advocated in the literature several times. Earlier examples include [Wan68] while more recent examples include: [SKA01, dAHM01, KA03] in the context of hybrid systems, [BMP02] in the context of quantized control systems, [FDF05] in the context of maneuver automata, [Lun94] in the context of stochastic models, [AMP95] in the context of discrete and timed-systems, among many other examples.

Part IV

Infinite Systems: Approximate symbolic models

9

Approximate system relationships

The similarity relationships introduced in Chapter 4 provided the framework upon which most of the abstraction techniques in Part III relied. In this chapter, we take an important conceptual step forward by abandoning the exact nature of these relationships.

Notation

A metric on a set Z is a function $\mathbf{d} : Z \times Z \to \mathbb{R}_0^+$ satisfying: $\mathbf{d}(z, z') = 0$ iff $z = z'$; $\mathbf{d}(z, z') + \mathbf{d}(z', z'') \geq \mathbf{d}(z, z'')$; $\mathbf{d}(z, z') = \mathbf{d}(z', z)$. A metric $\mathbf{d}$ on the set Z induces a distance between points $z \in Z$ and sets $W \subseteq Z$ by $\mathbf{d}(z, W) = \min_{w \in W} \mathbf{d}(z, w)$. This distance can be used to define the ε-inflation of a set $W \subseteq Z$, denoted by W^ε, and defined by $W^\varepsilon = \{z \in Z \mid \mathbf{d}(z, W) \leq \varepsilon\}$ for any $\varepsilon \in \mathbb{R}_0^+$. The set W^ε contains all the points in Z whose distance to W is bounded by ε. Note that $W \subset W^\varepsilon$ since $\mathbf{d}(w, W) = 0$ for any $w \in W$. Every relation $Q \subseteq Z \times W$, admits $Q^{-1} = \{(w, z) \in W \times Z \mid (z, w) \in Q\}$ as its inverse relation.

9.1 Approximate similarity relationships

The notion of simulation relation, formalized in Definition 4.7, requires related states to be sent by the output maps to the same output. It may be argued that such requirement is too strong since in concrete physical systems this exact equality is seldom achieved. Noise in measurements, imprecisions in actuators, and numerical computation errors are some of the factors preventing an exact equality between the outputs. These arguments suggest that one could relax the equality requirement by allowing related states to correspond to different outputs provided that the mismatch is bounded by some desired precision $\varepsilon \in \mathbb{R}_0^+$. To quantify the desired precision we need a metric on the set of outputs.

P. Tabuada, *Verification and Control of Hybrid Systems: A Symbolic Approach*,
DOI: 10.1007/978-1-4419-0224-5_9,

Definition 9.1 (Metric system). *A system S is said to be a* metric system *if the set of outputs Y is equipped with a metric $\mathbf{d}: Y \times Y \to \mathbb{R}_0^+$.*

When referring to metric systems, equality between two sets of outputs Y_a and Y_b will also imply equality between the corresponding metrics, *i.e.*, $Y_a = Y_b$ entails $\mathbf{d}_a = \mathbf{d}_b$ where $\mathbf{d}_a$ is the metric on Y_a and $\mathbf{d}_b$ is the metric on Y_b. For metric systems it is possible to generalize Definition 4.7 by replacing the second requirement with an approximate version.

Definition 9.2 (Approximate Simulation Relation). *Consider two metric systems S_a and S_b with $Y_a = Y_b$, and let $\varepsilon \in \mathbb{R}_0^+$. A relation $R \subseteq X_a \times X_b$ is an ε-*approximate simulation relation *from S_a to S_b if the following three conditions are satisfied:*

1. *for every $x_{a0} \in X_{a0}$, there exists $x_{b0} \in X_{b0}$ with $(x_{a0}, x_{b0}) \in R$;*
2. *for every $(x_a, x_b) \in R$ we have $\mathbf{d}(H_a(x_a), H_b(x_b)) \leq \varepsilon$;*
3. *for every $(x_a, x_b) \in R$ we have that:*
   $x_a \xrightarrow[a]{u_a} x'_a$ in S_a implies the existence of $x_b \xrightarrow[b]{u_b} x'_b$ in S_b satisfying $(x'_a, x'_b) \in R$.

We say that S_a is ε-approximately simulated by S_b or that S_b ε-approximately simulates S_a, denoted by $S_a \preceq_S^\varepsilon S_b$, if there exists an ε-approximate simulation relation from S_a to S_b.

When $\varepsilon = 0$ the inequality $\mathbf{d}(H_a(x_a), H_b(x_b)) \leq \varepsilon$ implies $H_a(x_a) = H_b(x_b)$. In this sense, we can regard approximate simulations as a generalization of the exact simulations introduced in Chapter 4. Before proceeding further, we give an example to illustrate this concept.

Example 9.3. Consider the dynamical system Σ described by the differential equation:

$$\frac{d}{dt}\xi = -\xi, \quad \xi(t) \in \mathbb{R}, t \in \mathbb{R}_0^+ \tag{9.1}$$

that can be explicitly integrated to obtain $\xi_x(t) = e^{-t}x$. The closed form expression for ξ is used to show that for any $\varepsilon \in \mathbb{R}_0^+$, the relation $R_\varepsilon \subseteq \mathbb{R} \times \mathbb{R}$ defined by $(x, x') \in R_\varepsilon$ iff $\|x - x'\| \leq \varepsilon$ is an ε-approximate simulation relation from $S(\Sigma)$ to $S(\Sigma)$. Here, $S(\Sigma)$ is the system $(\mathbb{R}, \mathbb{R}_0^+, \longrightarrow)$ defined by $x \xrightarrow{\tau} x'$ if there exists a solution $\xi_x : [0, \tau] \to \mathbb{R}$ of (9.1) satisfying $\xi_x(\tau) = x'$. To see why R_ε is an ε-approximate simulation relation, consider a pair $(x, x') \in R_\varepsilon$ and a transition $x \xrightarrow{\tau} x''$ in $S(\Sigma)$. The definition of $\longrightarrow$ implies $x'' = \xi_x(\tau) = e^{-\tau}x$, and we claim that $(x'', x''') \in R_\varepsilon$ with $x''' = \xi_{x'}(\tau)$, or equivalently, $x' \xrightarrow{\tau} x'''$ in $S(\Sigma)$. To determine if $(x'', x''') \in R_\varepsilon$, we compute:

$$\|x'' - x'''\| = \|\xi_x(\tau) - \xi_{x'}(\tau)\| = \|e^{-\tau}x - e^{-\tau}x'\| \leq \|e^{-\tau}\|\|x - x'\| \leq \|x - x'\| \leq \varepsilon.$$

This simple argument is valid in far greater generality and it is at the heart of all the results to be proved in Part IV. $\triangleleft$

As the previous example suggests, approximate simulation relations are especially useful for infinite-state systems in which the output set is naturally endowed with a metric. One typical usage of approximate simulation relations is the simplification of verification problems. To understand how such simplification arises, we relate the reachable sets of systems related by approximate simulation relations.

Proposition 9.4. *For any two metric systems S_a and S_b with $Y_a = Y_b$, the following implication holds:*

$$S_a \preceq_S^\varepsilon S_b \implies \mathrm{Reach}(S_a) \subseteq \mathrm{Reach}^\varepsilon(S_b).$$

Proof. Denote by R the ε-approximate simulation relation from S_a to S_b, and let $y_a \in \mathrm{Reach}(S_a)$. By definition of reachable output, there exists an initialized finite internal behavior of S_a:

$$x_{a0} \xrightarrow[a]{u_{a0}} x_{a1} \xrightarrow[a]{u_{a1}} \cdots \xrightarrow[a]{u_{ak-1}} x_{ak}$$

with $H_a(x_{ak}) = y_a$. Repeating the argument in the proof of Proposition 4.11 we conclude the existence of an initialized internal behavior of S_b:

$$x_{b0} \xrightarrow[b]{u_{b0}} x_{b1} \xrightarrow[b]{u_{b1}} \cdots \xrightarrow[b]{u_{bk-1}} x_{bk}$$

satisfying $(x_{ai}, x_{bi}) \in R$ for $i = 0, 1, \ldots, k$. Hence, $y_b = H_b(x_{bk}) \in \mathrm{Reach}(S_b)$ and it follows from the second requirement in the definition of approximate simulation relation that $\mathbf{d}(y_a, y_b) \leq \varepsilon$. Consequently, $y_a \in \mathrm{Reach}^\varepsilon(S_b)$. □

Returning to verification problems, consider a system S_a and a set of unsafe outputs B. If a system S_b ε-approximately simulates system S_a, Proposition 9.4 can be used to conclude that $\mathrm{Reach}^\varepsilon(S_b) \cap B = \varnothing$ implies $\mathrm{Reach}(S_a) \cap B = \varnothing$. For reachability problems, showing $\mathrm{Reach}(S_a) \cap Z \neq \varnothing$ for a set Z satisfying $Z^\varepsilon \subseteq B$, implies $\mathrm{Reach}(S_b) \cap B \neq \varnothing$. Clearly, these implications are only useful if we are able to construct abstractions based on ε-approximate simulation relations that are simpler than the systems they abstract. In Chapters 10 and 11 we discuss the existence and construction of such abstractions. Before, however, we strengthen approximate simulation to approximate bisimulation.

Definition 9.5 (Approximate bisimulation). *Consider two metric systems S_a and S_b with $Y_a = Y_b$, and let $\varepsilon \in \mathbb{R}_0^+$. We say that system S_a is* ε-approximately bisimilar *to system S_b, denoted by $S_a \cong_S^\varepsilon S_b$, if there exists a relation R satisfying:*

1. *R is an ε-approximate simulation relation from S_a to S_b;*
2. *R^{-1} is an ε-approximate simulation relation from S_b to S_a.*

Some care needs to be exerted when composing approximate (bi)simulation relations. Although it is a simple exercise to show that the composition of approximate (bi)simulation relations results in an approximate (bi)simulation relation, the precision is altered by composition. In detail, if ${}_aR_b$ is an ${}_a\varepsilon_b$-approximate (bi)simulation relation from S_a to S_b and if ${}_bR_c$ is an ${}_b\varepsilon_c$-approximate (bi)simulation relation from S_b to S_c, the composite ${}_bR_c \circ {}_aR_b$ is an $({}_a\varepsilon_b + {}_b\varepsilon_c)$-approximate (bi)simulation relation from S_a to S_c.

9.2 Approximate alternating similarity relationships

When discussing problems of control, ε-approximate similarity relationships need to be replaced with ε-approximate alternating similarity relationships. This generalization from exact to approximate consists again in relaxing the equality requirement on the outputs of related states.

Definition 9.6 (Approximate alternating simulation relation). *Let S_a and S_b be metric systems with $Y_a = Y_b$ and let $\varepsilon \in \mathbb{R}_0^+$. A relation $R \subseteq X_a \times X_b$ is an* ε-approximate alternating simulation relation *from S_a to S_b if the following three conditions are satisfied:*

1. *for every $x_{a0} \in X_{a0}$ there exists $x_{b0} \in X_{b0}$ with $(x_{a0}, x_{b0}) \in R$;*
2. *for every $(x_a, x_b) \in R$ we have $\mathbf{d}(H_a(x_a), H_b(x_b)) \leq \varepsilon$;*
3. *for every $(x_a, x_b) \in R$ and for every $u_a \in U_a(x_a)$ there exists $u_b \in U_b(x_b)$ such that for every $x'_b \in \mathrm{Post}_{u_b}(x_b)$ there exists $x'_a \in \mathrm{Post}_{u_a}(x_a)$ satisfying $(x'_a, x'_b) \in R$.*

We say that S_a is ε-approximately alternatingly simulated by S_b or that S_b ε-approximately alternatingly simulates S_a, denoted by $S_a \preceq^{\varepsilon}_{\mathcal{AS}} S_b$, if there exists an ε-approximate alternating simulation relation from S_a to S_b.

Approximate alternating simulation relations are used to define approximate feedback composition in Chapter 11. To that purpose, we introduce now the extended ε-approximate alternating simulation relation associated with an ε-approximate alternating simulation relation.

Definition 9.7 (Extended approximate alternating simulation relation). *Let R be an ε-approximate alternating simulation relation from metric system S_a to metric system S_b. The* extended ε-approximate alternating simulation relation *$R^e \subseteq X_a \times X_b \times U_a \times U_b$ associated with R is defined by all the quadruples $(x_a, x_b, u_a, u_b) \in X_a \times X_b \times U_a \times U_b$ for which the following three conditions hold:*

1. *$(x_a, x_b) \in R$;*
2. *$u_a \in U_a(x_a)$;*
3. *$u_b \in U_b(x_b)$ and for every $x'_b \in \mathrm{Post}_{u_b}(x_b)$ there exists $x'_a \in \mathrm{Post}_{u_a}(x_a)$ satisfying $(x'_a, x'_b) \in R$.*

Note that the third requirement in the previous definition is no more than the third requirement in Definition 9.6.

Approximate alternating bisimulations can be obtained by introducing the adjective approximate in the definition of alternating bisimulation or by symmetrizing the definition of approximate alternating simulation.

Definition 9.8 (Approximate alternating bisimulation). *Given two metric systems S_a and S_b with $Y_a = Y_b$, and given $\varepsilon \in \mathbb{R}_0^+$, we say that S_a is ε-approximately alternatingly bisimilar to S_b, denoted by $S_a \cong_{\mathcal{AS}}^{\varepsilon} S_b$, if there exists a relation R satisfying:*

1. *R is an ε-approximate alternating simulation relation from S_a to S_b;*
2. *R^{-1} is an ε-approximate alternating simulation relation from S_b to S_a.*

Approximate alternating simulations and bisimulations are instrumental to refine controllers synthesized for symbolic abstractions based on approximate simulations and bisimulations. We return to this topic in Chapter 11.

9.3 Notes

Approximate equivalence was first discussed in the context of timed-automata [GHJ97] and probabilistic systems [DGJP99]. In both cases, it was formalized by resorting to metrics and metric systems. Although in a different context, metric systems had been studied much earlier, see for example [vB98]. Most of the work that followed the papers [GHJ97, DGJP99] focused on probabilistic systems and the notion of approximate simulation for dynamical and control systems only appeared recently. In [GP05, GP07], approximate bisimulation was introduced by resorting to a metric on the set of outputs. A different formalization of approximate simulation appeared in [Tab05, Tab06] through the use of set-valued output maps. The discussion in this chapter is based on [GP07, PGT08].

10

Approximate symbolic models for verification

The abstraction techniques presented in Chapter 7 and Chapter 8 were based on the construction of quotient systems. In generalizing exact to approximate similarity relationships, we abandon quotient based abstractions to focus on a different abstraction technique introduced in this chapter for affine dynamical systems. Similar results for nonlinear dynamical systems are presented as special topics. The results in Chapter 11 further enlarge the class of approximate symbolic models for verification by considering the effect of adversarial inputs.

Notation

For any matrix $P \in \mathbb{R}^{n \times n}$, P^T denotes the transposed matrix. Matrix P is said to be symmetric if $P^T = P$, and is said to be positive definite if for every $x \in \mathbb{R}^n$, $x \neq 0$ implies $x^T P x > 0$. We denote by $\mathcal{SP}(n)$ the set of all symmetric and positive definite matrices in $\mathbb{R}^{n \times n}$. The minimum and the maximum eigenvalues of a matrix $P \in \mathbb{R}^{n \times n}$ are denoted by $\lambda_m(P)$ and $\lambda_M(P)$, respectively. For any $x \in \mathbb{R}^n$, $\|x\|$ represents the Euclidean norm of x defined by $\|x\| = \left(x_1^2 + x_2^2 + \ldots + x_n^2\right)^{\frac{1}{2}}$ where x_i is the ith component of the vector x. This norm induces a norm in the space of matrices that can be computed as $\|A\| = \lambda_M^{\frac{1}{2}}(A^T A)$ for any $A \in \mathbb{R}^{n \times m}$. The exponential of any matrix $A \in \mathbb{R}^{n \times n}$ is denoted by e^A and is the analytic function $\sum_{i=0}^{\infty} \frac{1}{i!} A^i$. The ball of radius $r \in \mathbb{R}_0^+$ centered at $x \in \mathbb{R}^n$ is denoted by $\mathcal{B}_r(x)$ and defined as the set of all the points $x' \in \mathbb{R}^n$ satisfying $\|x - x'\| \leq r$. If $Z \subseteq \mathbb{R}^n$ and $\eta \in \mathbb{R}^+$, $[Z]_\eta$ denotes the subset $[Z]_\eta \subseteq Z$ defined by:

$$[Z]_\eta = \left\{ z \in Z \mid z_i = k_i \frac{2}{\sqrt{n}} \eta \text{ for some } k_i \in \mathbb{Z} \text{ and } i = 1, 2, \ldots, n \right\}.$$

Note that we can cover Z by balls of radius η centered at the points in $[Z]_\eta$. This observation is used several times in this chapter. Given a subset $W \subseteq Z$

P. Tabuada, *Verification and Control of Hybrid Systems: A Symbolic Approach*,
DOI: 10.1007/978-1-4419-0224-5_10,

we denote by $\imath : W \hookrightarrow Z$ the natural inclusion of W in Z taking $w \in W$ to $\imath(w) = w \in Z$. The identity map on Z is denoted by $1_Z : Z \to Z$.

10.1 Stability of linear dynamical systems

The results in this chapter require some classical notions of stability that are recalled in this section. In doing so, we freely use several concepts of dynamical systems introduced in Section 7.1.1. The starting point is an affine dynamical system described by a differential equation:

$$\frac{d}{dt}\xi = A\xi + h \tag{10.1}$$

with $\xi(t) \in \mathbb{R}^n$, $A \in \mathbb{R}^{n \times n}$, $h \in \mathbb{R}^n$, and $t \in \mathbb{R}$. The adjective affine qualifies the right hand side of (10.1) which is an affine function represented by the matrix A and the vector h. For this reason, we represent an affine dynamical system by the triple $\Sigma = (\mathbb{R}^n, A, h)$ or by the pair $\Sigma = (\mathbb{R}^n, A)$ when $h = 0$. In the later case we speak of a linear dynamical system. Recall that a solution ξ of (10.1) is given by:

$$\xi_x(\tau) = e^{A\tau}x + \int_0^\tau e^{A(\tau - t)}h\, dt. \tag{10.2}$$

The following discussion focuses on the linear case since affine systems can be handled by techniques similar to those developed for linear dynamical systems. An equilibrium point of a linear dynamical system is a point $x_e \in \mathbb{R}^n$ for which $Ax_e = 0$. The equality $Ax_e = 0$ implies that any solution ξ_{x_e} of (10.1) with $h = 0$ remains at x_e for all future time. The reader can easily check this observation by noting that $\xi_{x_e} = x_e$ is indeed a solution of (10.1) with $h = 0$ or by using (10.2). For linear dynamical systems, the origin is always an equilibrium point and we focus on this equilibrium point.

Many of the dynamical and hybrid dynamical systems to be verified arise as the result of applying a feedback control law to a physical system. In such cases, the objective for the continuous feedback control law is to render the equilibrium x_e asymptotically stable.

Definition 10.1 (Asymptotically stable equilibrium point). *The equilibrium point $x_e = 0$ of a linear dynamical system $\Sigma = (\mathbb{R}^n, A)$ is said to be* asymptotically stable *if there exist $\kappa, \lambda \in \mathbb{R}^+$ such that for every $x \in \mathbb{R}^n$ and every $t \in \mathbb{R}_0^+$, the following inequality is satisfied:*

$$\|\xi_x(t)\| \leq \kappa e^{-\lambda t}\|x\|. \tag{10.3}$$

When λ is allowed to become zero, the equilibrium point $x_e = 0$ is said to be stable.

Inequality (10.3) entails that the trajectory ξ_x is bounded for all time by $\kappa\|x\|$. Moreover, $\xi_x(t)$ converges asymptotically to the equilibrium $x_e = 0$, as $t \to \infty$, since:

$$\lim_{t\to\infty} \|\xi_x(t)\| \leq \lim_{t\to\infty} \kappa e^{-\lambda t}\|x\| = 0 \implies \lim_{t\to\infty} \xi_x(t) = 0.$$

Although (10.3) also states that ξ_x converges exponentially fast, we shall not make use of this fact. For linear dynamical systems, asymptotic stability can be checked by analyzing the eigenvalues of the matrix A.

Theorem 10.2. *The equilibrium point $x_e = 0$ of a linear dynamical system $\Sigma = (\mathbb{R}^n, A)$ is asymptotically stable iff all the eigenvalues of the matrix A have negative real part.*

For nonlinear dynamical systems we cannot use the preceding eigenvalue test and have to resort to an alternative criterion based on Lyapunov functions. This criterion is also useful for linear dynamical systems. In fact, all the results in this chapter are based on Lyapunov functions.

Definition 10.3 (Lyapunov function). *Let $\Sigma = (\mathbb{R}^n, A)$ be a linear dynamical system and consider a function $V : \mathbb{R}^n \to \mathbb{R}$ satisfying the following three properties:*

1. *V is continuous on $\mathbb{R}^n$ and smooth on $\mathbb{R}^n\backslash\{0\}$;*
2. *$V(x) \geq 0$ for all $x \in \mathbb{R}^n$;*
3. *$V(x) = 0$ implies $x = 0$.*

The function V is a Lyapunov function *for Σ if there exists $\lambda \in \mathbb{R}^+$ such that for every $x \in \mathbb{R}^n\backslash\{0\}$, the following inequality holds:*

$$\frac{\partial V}{\partial x} Ax \leq -\lambda V(x). \tag{10.4}$$

When λ is allowed to become zero, V is a weak Lyapunov function for Σ.

Inequality (10.4) is at the core of several results in this chapter. Note that if $\xi : \mathbb{R} \to \mathbb{R}^n$ is a trajectory of Σ, then $V \circ \xi : \mathbb{R} \to \mathbb{R}_0^+$ is a smooth function of time that obeys the following differential inequality:

$$\frac{d}{dt} V \circ \xi = \left.\frac{\partial V}{\partial x}\right|_{x=\xi} \frac{d\xi}{dt} = \left.\frac{\partial V}{\partial x}\right|_{x=\xi} A\xi \leq -\lambda V \circ \xi.$$

Integration, provides the bound:

$$V \circ \xi(t) \leq e^{-\lambda t} V(\xi(0))$$

showing that V decays exponentially along the trajectories of Σ. This observation can be used to show that Lyapunov functions completely characterize asymptotic stability for linear dynamical systems.

Theorem 10.4. *The equilibrium point $x_e = 0$ of a linear dynamical system Σ is asymptotically stable iff Σ admits a Lyapunov function. Analogously, the equilibrium point $x_e = 0$ of a linear dynamical system Σ is stable iff Σ admits a weak Lyapunov function.*

For linear dynamical systems, existence of a Lyapunov function also implies the existence of a Lyapunov function of the form $V(x) = \sqrt{x^T P x}$ for some $P \in \mathcal{SP}(n)$. Moreover, Lyapunov functions of this form satisfy several useful inequalities.

Proposition 10.5. *Let $V : \mathbb{R}^n \to \mathbb{R}_0^+$ be a function of the form $V(x) = \sqrt{x^T P x}$ for some $P \in \mathcal{SP}(n)$. There exist constants $\underline{\alpha}, \overline{\alpha}, \gamma \in \mathbb{R}^+$ such that for all $x, x', x'' \in \mathbb{R}^n$, the following inequalities are satisfied:*

$$\underline{\alpha}\|x\| \le V(x) \le \overline{\alpha}\|x\|, \tag{10.5}$$

$$V(x - x') - V(x - x'') \le \gamma\|x' - x''\|. \tag{10.6}$$

Proof. Let $V' = V^2$ and recall[1] that since $P = P^T$ and $P \in \mathbb{R}^{n \times n}$ there exists a matrix $Q \in \mathbb{R}^{n \times n}$ satisfying $Q^{-1} = Q^T$ and rendering the matrix $D = Q^{-1}PQ$ diagonal. Let now $z = Qx$ and note that:

$$V'(z) = V'(Qx) = x^T Q^T P Q x = x^T Q^{-1} P Q x = x^T D x.$$

Since D is diagonal and the eigenvalues of D and P are the same, we have:

$$\lambda_m(P)\|z\|^2 = \lambda_m(P) z^T z \le V'(z) \le \lambda_M(P) z^T z = \lambda_M(P)\|z\|^2.$$

Inequality (10.5) now follows by setting $\underline{\alpha} = \sqrt{\lambda_m(P)}$ and $\overline{\alpha} = \sqrt{\lambda_M(P)}$.

Fix $x \in \mathbb{R}^n$ and consider the function $g : \mathbb{R}^n \to \mathbb{R}_0^+$ defined by $g(z) = V(x - z)$. This function is differentiable for all $z \ne x$ and continuous for all $z \in \mathbb{R}^n$. Consider now two points $z, w \in \mathbb{R}^n$ and let L be line joining these two points. If $x \notin L$, then, by the mean value theorem, there exists a point $x^* \in L \subset \mathbb{R}^n$ that satisfies:

$$g(z) - g(w) = \left.\frac{\partial g}{\partial z}\right|_{z=x^*} (z - w).$$

The right hand side of the above equality is bounded by:

$$\left.\frac{\partial g}{\partial z}\right|_{z=x^*} (z - w) \le \left\| \left.\frac{\partial g}{\partial z}\right|_{z=x^*} (z - w) \right\| \le \left\| \left.\frac{\partial g}{\partial z}\right|_{z=x^*} \right\| \|z - w\|.$$

Moreover, $\|\partial g/\partial z\|$ is bounded by:

$$\left\| \frac{\partial g}{\partial z} \right\| = \left\| \frac{-2(x - z)^T P}{2V(x - z)} \right\| \le \frac{\|P\|\|x - z\|}{\sqrt{\lambda_m(P)}\|x - z\|} \le \frac{\lambda_M(P)}{\sqrt{\lambda_m(P)}}.$$

[1] See for example [AM97].

We now consider the case when $x \in L$. Since the points x, z, w belong to the same line L it follows that $x - w = \lambda(x - z)$ for some $\lambda \in \mathbb{R}_0^+$. Therefore:

$$\begin{aligned} V(x-z) - V(x-w) &= V(x-z) - V(\lambda(x-z)) \\ &= V(x-z) - \lambda V(x-z) \\ &= (1-\lambda)V(x-z) \\ &= V((1-\lambda)(x-z)) \\ &= V(x-z-\lambda(x-z)) \\ &= V(x-z-(x-w)) \\ &= V(z-w) \leq \sqrt{\lambda_M(P)}\|z-w\|. \end{aligned}$$

The result now follows by taking $\gamma = \max\left\{\frac{\lambda_M(P)}{\sqrt{\lambda_m(P)}}, \sqrt{\lambda_M(P)}\right\} = \frac{\lambda_M(P)}{\sqrt{\lambda_m(P)}}$.

□

The bounds given in (10.5) allow us to estimate the norm of a vector $x \in \mathbb{R}^n$ from the knowledge of $V(x)$ and vice-versa. Inequality (10.6) can be seen as a generalization of the triangular inequality. Recall that $\|x - x'\| \leq \|x - x''\| + \|x' - x''\|$ and thus $\|x - x'\| - \|x - x''\| \leq \|x' - x''\|$. Therefore, if $V(x - x') = \|x - x'\|$, (10.6) is satisfied with $\gamma = 1$.

10.2 Dynamical systems as systems

Dynamical systems can be modeled as systems in several different ways. In Section 7.1 we presented one construction based on a dynamical system and a finite equivalence relation defining the outputs. Here, we present a different construction where the set of outputs remains infinite so as to retain the Euclidean metric on $\mathbb{R}^n$.

Definition 10.6. *The system $S_\tau(\Sigma) = (X_\tau, U_\tau, \underset{\tau}{\longrightarrow}, Y_\tau, H_\tau)$ associated with a dynamical system $\Sigma = (\mathbb{R}^n, f)$ and with $\tau \in \mathbb{R}^+$ consists of:*

- $X_\tau = \mathbb{R}^n$;
- $U_\tau = \{\tau\}$;
- $x \overset{\tau}{\underset{\tau}{\longrightarrow}} x'$ *if there exists a solution* $\xi_x : [0, \tau] \to \mathbb{R}^n$ *of* Σ *satisfying* $\xi_x(\tau) = x'$;
- $Y_\tau = \mathbb{R}^n$;
- $H_\tau = \imath : X_\tau \to \mathbb{R}^n$.

System $S_\tau(\Sigma)$ only describes the states reached by trajectories of Σ of duration τ. We can thus regard $S_\tau(\Sigma)$ as a time-triggered sampled version of Σ where as the system $S_Q(\Sigma)$, described in Section 7.1, can be regarded as an output-triggered sampled version of Σ. The parameter τ defines the desired sampling rate and is a measure of time quantization. It is not difficult to see that $S_\tau(\Sigma)$ is a metric system since $Y = \mathbb{R}^n$ is equipped with the Euclidean metric $\mathbf{d}(y, y') = \|y - y'\|$.

10.3 Symbolic models for affine dynamical systems

The abstractions considered in this chapter are obtained by quantizing the state set of $S_\tau(\Sigma)$ and approximating its transitions.

Definition 10.7. *The system $S_{\tau\eta}(\Sigma) = (X_{\tau\eta}, U_{\tau\eta}, \underset{\tau\eta}{\longrightarrow}, Y_{\tau\eta}, H_{\tau\eta})$ associated with a dynamical system $\Sigma = (\mathbb{R}^n, f)$ and with $\tau, \eta \in \mathbb{R}^+$ consists of:*

- $X_{\tau\eta} = [\mathbb{R}^n]_\eta$;
- $U_{\tau\eta} = \{\tau\}$
- $x \overset{\tau}{\underset{\tau\eta}{\longrightarrow}} x'$ *if there exists a solution* $\xi_x : [0,\tau] \to \mathbb{R}^n$ *of* Σ *satisfying* $\|\xi_x(\tau) - x'\| \leq \eta$;
- $Y_{\tau\eta} = \mathbb{R}^n$;
- $H_{\tau\eta} = \imath : X_{\tau\eta} \hookrightarrow \mathbb{R}^n$.

Note that $S_{\tau\eta}(\Sigma)$ has countably many states and when the domain of f is a bounded set, $S_{\tau\eta}(\Sigma)$ becomes finite-state. Bounded domains occur frequently in applications where physical variables, such as temperatures and pressures, cannot grow unbounded under normal operating circumstances. Although $[\mathbb{R}^n]_\eta$ has the desirable consequence of rendering $X_{\tau\eta}$ countable, it also forces the transition relation of $S_{\tau\eta}(\Sigma)$ to inaccurately describe the transitions in $S_\tau(\Sigma)$ since $x \overset{\tau}{\underset{\tau\eta}{\longrightarrow}} x'$ does not entail $x' = \xi_x(\tau)$, but only $\|x' - \xi_x(\tau)\| \leq \eta$. The extent to which the inaccuracies in $\underset{\tau}{\longrightarrow}$ can be tolerated is described by the next result.

Theorem 10.8. *Consider a linear dynamical system Σ admitting a Lyapunov function V of the form $V(x) = \sqrt{x^T P x}$ with $P \in \mathcal{SP}(n)$. For any desired precision $\varepsilon \in \mathbb{R}^+$, for any desired time quantization $\tau \in \mathbb{R}^+$, and for any space quantization $\eta \in \mathbb{R}^+$ satisfying:*

$$\eta \leq \min\left\{\gamma^{-1}\underline{\alpha}\varepsilon\left(1 - e^{-\lambda\tau}\right), \overline{\alpha}^{-1}\underline{\alpha}\varepsilon\right\}, \qquad (10.7)$$

the relation $R_\varepsilon \subseteq X_\tau \times X_{\tau\eta}$ defined by:

$$R_\varepsilon = \{(x_\tau, x_{\tau\eta}) \in X_\tau \times X_{\tau\eta} \mid V(x_\tau - x_{\tau\eta}) \leq \underline{\alpha}\varepsilon\} \qquad (10.8)$$

is an ε-approximate bisimulation relation between $S_\tau(\Sigma)$ and $S_{\tau\eta}(\Sigma)$.

Proof. The proof consists in showing that the relation R_ε satisfies all the conditions in the definition of ε-approximate bisimulation relation. We first show that R_ε is an ε-approximate simulation relation from $S_\tau(\Sigma)$ to $S_{\tau\eta}(\Sigma)$.

The first requirement in Definition 9.2 asks that for every $x_{\tau 0} \in X_{\tau 0} = X_\tau$ there exists $x_{\tau\eta 0} \in X_{\tau\eta 0} = X_{\tau\eta}$ satisfying $(x_{\tau 0}, x_{\tau\eta 0}) \in R_\varepsilon$ or equivalently

$V(x_{\tau 0} - x_{\tau\eta 0}) \leq \underline{\alpha}\varepsilon$. If $x_\tau \in X_{\tau 0}$ then, by definition of $X_{\tau\eta 0}$, there exists $x_{\tau\eta 0} \in X_{\tau\eta 0}$ satisfying $\|x_{\tau 0} - x_{\tau\eta 0}\| \leq \eta$. Consequently:

$$V(x_{\tau 0} - x_{\tau\eta 0}) \leq \overline{\alpha}\|x_{\tau 0} - x_{\tau\eta 0}\| \leq \overline{\alpha}\eta \leq \underline{\alpha}\varepsilon \tag{10.9}$$

where the first inequality follows from (10.5) and the second follows from (10.7).

The second requirement holds by construction since $(x_\tau, x_{\tau\eta}) \in R_\varepsilon$, (10.5), and (10.8) imply:

$$\mathbf{d}(H_\tau(x_\tau), H_{\tau\eta}(x_{\tau\eta})) = \|x_\tau - x_{\tau\eta}\| \leq \frac{1}{\underline{\alpha}} V(x_\tau - x_{\tau\eta}) \leq \varepsilon.$$

Regarding the third requirement, let $(x_\tau, x_{\tau\eta}) \in R_\varepsilon$ and consider the transition $x_\tau \xrightarrow[\tau]{\tau} x'_\tau$ in $S_\tau(\Sigma)$. Since the origin is an asymptotically stable equilibrium point, Σ is forward complete and the transition $x_{\tau\eta} \xrightarrow[\tau]{\tau} x''_{\tau\eta}$ is well defined. By definition of $S_{\tau\eta}(\Sigma)$, for any $x'_{\tau\eta} \in X_{\tau\eta}$ satisfying:

$$\|x''_{\tau\eta} - x'_{\tau\eta}\| \leq \eta, \tag{10.10}$$

$x_{\tau\eta} \xrightarrow[\tau\eta]{\tau} x'_{\tau\eta}$ is a transition in $S_{\tau\eta}(\Sigma)$. Note that a point $x'_{\tau\eta}$ satisfying (10.10) always exists since $\mathbb{R}^n \subseteq \cup_{z \in X_{\tau\eta}} \mathcal{B}_\eta(z)$. We now need to show that $(x'_\tau, x'_{\tau\eta})$ belongs to R_ε, *i.e.*, $V(x'_\tau - x'_{\tau\eta}) \leq \underline{\alpha}\varepsilon$. This follows from the sequence of inequalities:

$$\begin{aligned}
V(x'_\tau - x'_{\tau\eta}) &\leq V(x'_\tau - x''_{\tau\eta}) + \gamma\|x'_{\tau\eta} - x''_{\tau\eta}\| && (10.11)\\
&\leq V(\xi_{x_\tau}(\tau) - \xi_{x_{\tau\eta}}(\tau)) + \gamma\eta && (10.12)\\
&\leq e^{-\lambda\tau} V(\xi_{x_\tau}(0) - \xi_{x_{\tau\eta}}(0)) + \gamma\eta && (10.13)\\
&\leq e^{-\lambda\tau}\underline{\alpha}\varepsilon + \gamma\eta && (10.14)\\
&\leq \underline{\alpha}\varepsilon && (10.15)
\end{aligned}$$

where we used (10.6), (10.10), (10.8), and (10.7) in the first, second, fourth, and fifth inequalities, respectively. The third inequality is a consequence of:

$$V(\xi_x(\tau) - \xi_{x'}(\tau)) \leq e^{-\lambda\tau} V(\xi_x(0) - \xi_{x'}(0)), \tag{10.16}$$

proved in the slightly more general context of Corollary 10.10.

The proof that R_ε^{-1} is an ε-approximate simulation relation from $S_{\tau\eta}(\Sigma)$ to $S_\tau(\Sigma)$ is similar and, for that reason, omitted. □

The main assumption in Theorem 10.8 is asymptotic stability of the origin. This assumption is satisfied by a large class of verification problems where the dynamical system being verified results from applying a feedback control law, to a physical system, with the objective of achieving asymptotic stability. The inequality (10.7) describes the tradeoff between the desired precision ε, the desired time quantization τ, and the required space quantization η.

The construction of $S_{\tau\eta}(\Sigma)$ is not based on an equivalence relation and its corresponding quotient, but rather on a judicious approximation of infinitely many transitions in $S_\tau(\Sigma)$. In particular, a transition $x \xrightarrow[\tau\eta]{\tau} x''$ in $S_{\tau\eta}(\Sigma)$ represents the infinitely many transitions $x' \xrightarrow[\tau]{\tau} x'''$ in $S_\tau(\Sigma)$ with $V(x - x') \leq \underline{\alpha}\varepsilon$. This approximation is possible due to the stability properties of the class of differential equations being considered and the flexibility afforded by the approximate nature of approximate bisimulation relations.

Example 10.9. We recall here the example used to illustrate the exact abstraction techniques discussed in Chapter 7. It consists of the linear dynamical system defined by:

$$\dot{\xi}_1 = -7\xi_1 + \xi_2$$
$$\dot{\xi}_1 = 8\xi_1 - 10\xi_2,$$

the set of initial states:

$$L = \{(x_1, x_2) \in \mathbb{R}^2 \mid 5 \leq x_1 \leq 6 \ \wedge \ -1 \leq x_2 \leq 1\},$$

and the set of unsafe states:

$$B = \{(x_1, x_2) \in \mathbb{R}^2 \mid x_2 \leq -4 \ \vee \ x_2 \geq 4\}.$$

The objective is to determine if the trajectories of Σ starting in L avoid the set B. In order to apply Theorem 10.8 we identify $V(x) = \sqrt{x_1^2 + x_2^2}$ as a suitable Lyapunov function. We have:

$$P = I_2, \quad \underline{\alpha} = 1, \quad \overline{\alpha} = 1, \quad \lambda = \frac{7}{2},$$

where I_2 is the 2×2 identity matrix. For this choice of Lyapunov function we can directly verify that $\gamma = 1$:

$$\begin{aligned} V(x - x') - V(x - x'') &\leq \overline{\alpha}\|x - x'\| - \underline{\alpha}\|x - x''\| \\ &= \|x - x'\| - \|x - x''\| \\ &\leq \|x' - x''\|. \end{aligned}$$

Choosing $\varepsilon = 0.5$ and $\tau = 0.05$, inequality (10.7) requires $\eta \leq \min\{0.08027, 1\}$. We choose $\eta = \frac{\sqrt{2}}{20} \approx 0.0707$ thus satisfying (10.7). Using these parameters and restricting Σ to the bounded set:

$$\{(x_1, x_2) \in \mathbb{R}^2 \mid -0.5 \leq x_1 \leq 6.5 \ \wedge \ -1.5 \leq x_2 \leq 4.5\},$$

we compute, through numerical simulation, the finite-state bisimilar system $S_{\tau\eta}(\Sigma)$. In Figure (10.1) the reader can appreciate: dots representing the states $X_{\tau\eta}$ of $S_{\tau\eta}(\Sigma)$; larger and dark-colored dots representing $X_{\tau\eta} \cap B$; and

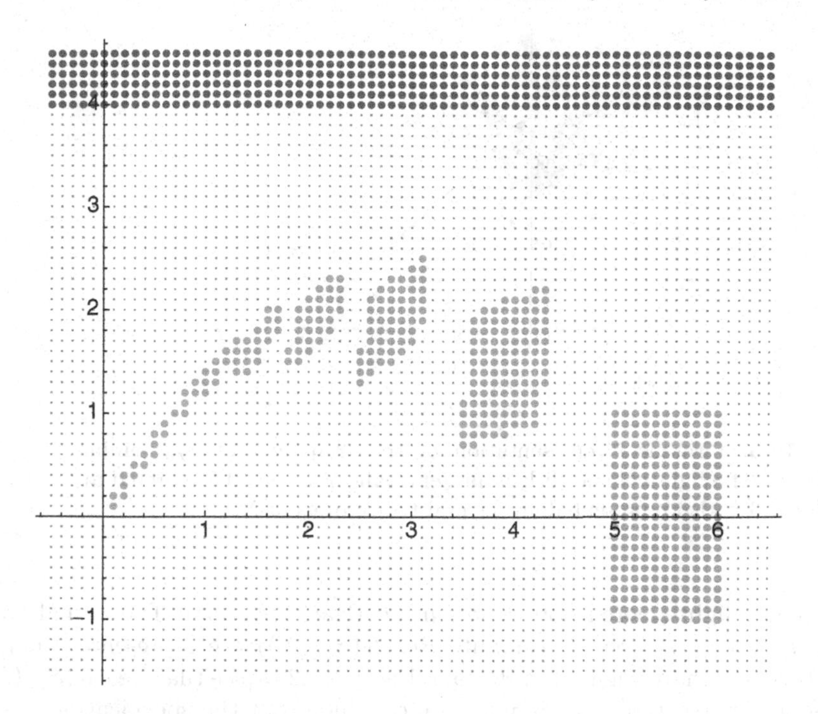

Fig. 10.1. Graphical representation of the states reachable from L. The dots represent the states of $S_{\tau\eta}(\Sigma)$, the larger and dark-colored dots represent $X_{\tau\eta} \cap B$ while the larger and light-colored dots represent $\mathrm{Reach}\,(S_{\tau\eta}(\Sigma))$.

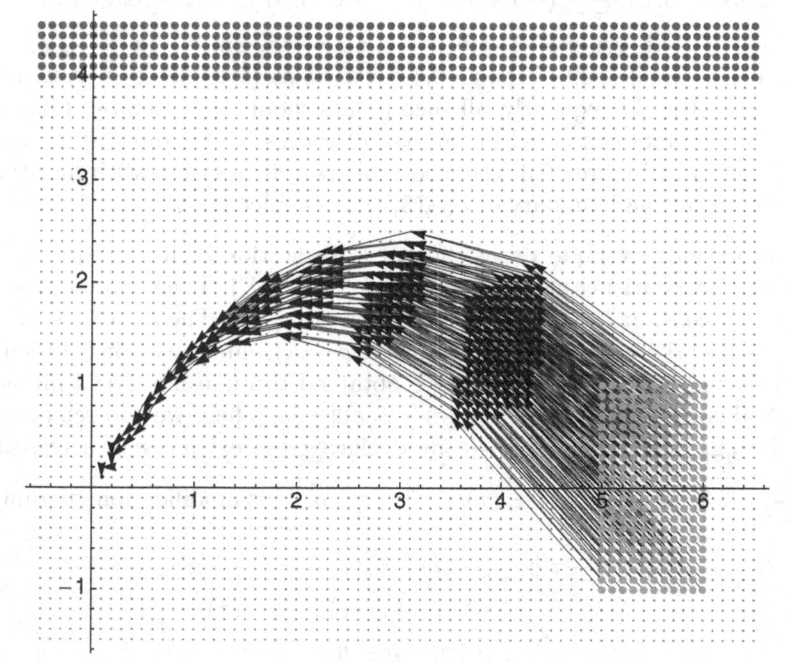

Fig. 10.2. Graphical representation of the transitions of $S_{\tau\eta}(\Sigma)$ used to compute the states reachable from L displayed in Figure 10.1.

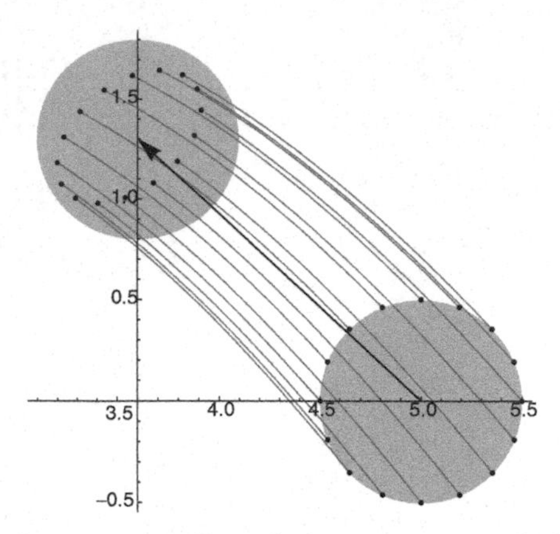

Fig. 10.3. Graphical representation of the transition of $S_{\tau\eta}$ joining $(5,0)$ to $(3.6,1.3)$. The light-colored disks represent balls of radius 0.5 centered at $(5,0)$ and at $(3.6,1.3)$. Also represented are several trajectories of Σ.

larger and light-colored dots representing Reach $(S_{\tau\eta}(\Sigma))$. The actual transitions of $S_{\tau\eta}(\Sigma)$ used to compute Reach $(S_{\tau\eta}(\Sigma))$ are represented in Figure 10.2. By analyzing Figure 10.1 and Figure 10.2 we see that Reach $(S_{\tau\eta}(\Sigma))$ does not intersect $B^{0.5}$ from which we conclude that the specification is satisfied.

To illustrate the approximate nature of $S_{\tau\eta}(\Sigma)$, we show in Figure 10.3 the transition $(5,0) \underset{\tau\eta}{\longrightarrow} (3.6,1.3)$. The state $(5,0) \in X_{\tau\eta}$ is related to all the states $x \in X_\tau$ satisfying $V(x-(5,0)) \leq \underline{\alpha}\varepsilon = 0.5$. Since $V(x) = \sqrt{x_1^2 + x_2^2}$, all the points in the ball of radius 0.5 centered at $(5,0)$ are ε-approximately related to $(5,0)$. Consequently, all such points should be taken by transitions of $S_\tau(\Sigma)$ to points contained in the ball of radius 0.5 centered at $(3.6,1.3)$. This is shown in Figure 10.3 where several trajectories of Σ with duration τ and starting on the boundary of $\mathcal{B}_{0.5}(5,0)$ are depicted. $\lhd$

The previous example raises the natural question of how to handle the numerical errors arising in the computation of $S_{\tau\eta}(\Sigma)$. These can be incorporated in the construction of $S_{\tau\eta}(\Sigma)$ while ensuring that Theorem 10.8 remains valid. Assume the existence of a parameter ρ describing the error between the true trajectory ξ and the trajectory $\widehat{\xi}$ obtained through numerical methods, when evaluated at time τ, *i.e.*, $\|\xi(\tau) - \widehat{\xi}(\tau)\| \leq \rho$. Then, we can replace $X_{\tau\eta}$ with $[\mathbb{R}^n]_{\eta-\rho}$ and redefine $\underset{\tau\eta}{\longrightarrow}$ by declaring the existence of a transition $x \overset{\tau}{\underset{\tau\eta}{\longrightarrow}} x'$ when $\widehat{\xi}_x$ satisfies $\|\widehat{\xi}_x(\tau) - x'\| \leq \eta - \rho$. This modification guarantees:

$$\begin{aligned}
\|\xi_x(\tau) - x'\| &\leq \|\xi_x(\tau) - \widehat{\xi}_x(\tau) + \widehat{\xi}_x(\tau) - x'\| \\
&\leq \|\xi_x(\tau) - \widehat{\xi}_x(\tau)\| + \|\widehat{\xi}_x(\tau) - x'\| \\
&\leq \rho + \eta - \rho = \eta
\end{aligned}$$

which is the original definition of $\underset{\tau\eta}{\longrightarrow}$ in the absence of numerical errors.

A different concern that may arise when using $S_{\tau\eta}(\Sigma)$ for verification is the inter-sample behavior: can a specification be violated for $t \in]0, \tau[$ even though it is satisfied at $t = 0$ and $t = \tau$? The techniques employed in Section 7.6 for the computation of reachable sets provide an answer to this concern. Recall that a transition $x_{\tau\eta} \overset{\tau}{\underset{\tau\eta}{\longrightarrow}} x'_{\tau\eta}$ in $S_{\tau\eta}$ implies the existence of a trajectory $\xi_{x_{\tau\eta}}$ of Σ satisfying $\|x'_{\tau\eta} - \xi_{x_{\tau\eta}}(\tau)\| \leq \eta$. We can thus enclose $x_{\tau\eta}$ in a zonotope, enclose $\mathcal{B}_\eta(x'_{\tau\eta})$ in a different zonotope, and use Proposition 7.31 to obtain another zonotope containing all the states $\xi_{x_{\tau\eta}}(t)$ for $t \in [0, \tau]$. In practice, however, the parameter τ is chosen to be sufficiently small so that the specification is directly verified against $S_{\tau\eta}(\Sigma)$.

Theorem 10.8 admits a simple generalization to affine dynamical systems.

Corollary 10.10. *Consider an affine dynamical system $\Sigma = (\mathbb{R}^n, A, h)$ and assume that the linear dynamical system $(\mathbb{R}^n, A)$ admits a Lyapunov function V of the form $V(x) = \sqrt{x^T P x}$ with $P \in \mathcal{SP}(n)$. For any desired precision $\varepsilon \in \mathbb{R}^+$, for any desired time quantization $\tau \in \mathbb{R}^+$, and for any space quantization $\eta \in \mathbb{R}^+$ satisfying:*

$$\eta \leq \min \left\{ \gamma^{-1} \underline{\alpha} \varepsilon \left(1 - e^{-\lambda\tau}\right), \overline{\alpha}^{-1} \underline{\alpha} \varepsilon \right\}, \tag{10.17}$$

the relation $R_\varepsilon \subseteq X_\tau \times X_{\tau\eta}$ defined by:

$$R_\varepsilon = \{(x_\tau, x_{\tau\eta}) \in X_\tau \times X_{\tau\eta} \mid V(x_\tau - x_{\tau\eta}) \leq \underline{\alpha}\varepsilon\} \tag{10.18}$$

is an ε-approximate bisimulation relation between $S_\tau(\Sigma)$ and $S_{\tau\eta}(\Sigma)$.

Proof. The proof is the same as the proof of Theorem 10.8 except for a simple modification needed to show that (10.16) holds for all affine dynamical systems. For completeness we now present this argument. Let $x', x'' \in \mathbb{R}^n$. Considering $V(\xi_{x'}(t) - \xi_{x''}(t))$ as a function of time we have:

$$\begin{aligned}
\frac{d}{dt} V(\xi_{x'} - \xi_{x''}) &= \left. \frac{\partial V}{\partial x} \right|_{x = \xi_{x'} - \xi_{x''}} \frac{d}{dt} (\xi_{x'} - \xi_{x''}) \\
&= \left. \frac{\partial V}{\partial x} \right|_{x = \xi_{x'} - \xi_{x''}} (A\xi_{x'} + h - A\xi_{x''} - h) \\
&= \left. \frac{\partial V}{\partial x} \right|_{x = \xi_{x'} - \xi_{x''}} A(\xi_{x'} - \xi_{x''}) \\
&\leq -\lambda V(\xi_{x'} - \xi_{x''}).
\end{aligned}$$

By integration we obtain:

$$V(\xi_{x'}(t) - \xi_{x''}(t)) \leq e^{-\lambda t} V(\xi_{x'}(0) - \xi_{x''}(0))$$

which is the desired inequality. □

Theorem 10.8 and Corollary 10.10 rely on the existence of a Lyapunov function. From Theorem 10.4 we know that existence of a Lyapunov function is equivalent to asymptotic stability of the origin. Therefore, we can first check for existence of a Lyapunov function by determining if the origin is an asymptotically stable equilibrium point. According to Theorem 10.2 this can be done by determining if the eigenvalues of the matrix A have negative real part. When this is the case, we can resort to one of several well established methods in control theory to compute the matrix P defining the required Lyapunov function.

Lyapunov functions are more than a convenient tool in the construction of symbolic abstractions. Under the assumptions of Theorem 10.8, the relation defined by the pairs $(x_\tau, x'_\tau) \in X_\tau \times X_\tau$ satisfying $V(x_\tau - x'_\tau) \leq \underline{\alpha}\varepsilon$ is in fact an ε-approximate bisimulation relation between $S_\tau(\Sigma)$ and $S_\tau(\Sigma)$ for any $\tau \in \mathbb{R}^+$. This is consistent with what happens in the exact case where Theorem 4.18 tells us that existence of a bisimulation relation R between a system S and itself leads to a bisimulation relation between S and the symbolic abstraction $S_{/R}$. Although the abstractions considered in this chapter are not quotient based, they require identical assumptions on the system S to be abstracted: existence of an approximate bisimulation relation between S and S. Note also that (10.18) defines not one, but a family of approximate bisimulation relations parameterized by ε. The converse is also true under one additional assumption.

Proposition 10.11. *Consider an affine dynamical system* $\Sigma = (\mathbb{R}^n, A, h)$ *and let* V *be a weak Lyapunov function for the linear dynamical system* $(\mathbb{R}^n, A)$ *of the form* $V(x) = \sqrt{x^T P x}$ *with* $P \in \mathcal{SP}(n)$. *For any* $\varepsilon \in \mathbb{R}^+$ *and any* $\tau \in \mathbb{R}^+$, *the relation defined by:*

$$R_\varepsilon = \{(x_\tau, x'_\tau) \in X_\tau \times X_\tau \mid V(x_\tau - x'_\tau) \leq \underline{\alpha}\varepsilon\}$$

is an ε*-approximate bisimulation relation between* $S_\tau(\Sigma)$ *and* $S_\tau(\Sigma)$. *Conversely, assume the existence of* $\alpha \in \mathbb{R}^+$ *such that for any* $\varepsilon \in \mathbb{R}^+$ *and any* $\tau \in \mathbb{R}^+$, R_ε *is an* ε*-approximate bisimulation relation between* $S_\tau(\Sigma)$ *and* $S_\tau(\Sigma)$ *satisfying:*

$$\|x_\tau - x'_\tau\| \leq \alpha\varepsilon \implies (x_\tau, x'_\tau) \in R_\varepsilon. \tag{10.19}$$

Then, there exists a weak Lyapunov function V *for the linear dynamical system* $(\mathbb{R}^n, A)$.

Proof. We first note that since $R_\varepsilon^{-1} = R_\varepsilon$, it suffices to show that R_ε is an ε-approximate simulation relation from $S_\tau(\Sigma)$ to $S_\tau(\Sigma)$ to conclude that R_ε is also an ε-approximate bisimulation relation between $S_\tau(\Sigma)$ and $S_\tau(\Sigma)$. The first requirement in Definition 9.2 is immediate since $X_{\tau 0} = X_\tau$ and for any

$x_\tau \in X_\tau$, $(x_\tau, x_\tau) \in R_\varepsilon$. The second requirement follows at once from (10.5) and:

$$\mathbf{d}(H_\tau(x_\tau), H_\tau(x'_\tau)) = \|x_\tau - x'_\tau\| \leq \frac{1}{\underline{\alpha}} V(x_\tau - x'_\tau) \leq \varepsilon.$$

Regarding the third requirement, let $(x_\tau, x'_\tau) \in R_\varepsilon$ and $x_\tau \xrightarrow[\tau]{\tau} x''_\tau$ in $S_\tau(\Sigma)$. It suffices to show that $(x''_\tau, x'''_\tau) \in R_\varepsilon$ for $x'_\tau \xrightarrow[\tau]{\tau} x'''_\tau$. This follows from:

$$V(x''_\tau - x'''_\tau) = V(\xi_{x_\tau}(\tau) - \xi_{x'_\tau}(\tau)) \leq e^{-\lambda\tau} V(\xi_{x_\tau}(0) - \xi_{x'_\tau}(0)) \leq V(x_\tau - x'_\tau) \leq \underline{\alpha}\varepsilon$$

where we used (10.16).

Assume now the existence of $\alpha \in \mathbb{R}^+$ such that for any $\varepsilon \in \mathbb{R}^+$ and any $\tau \in \mathbb{R}^+$, R_ε is an ε-approximate bisimulation relation between $S_\tau(\Sigma)$ and $S_\tau(\Sigma)$ satisfying (10.19). Consider any point $x \in \mathbb{R}^n$ and let ε be $\|x\|/\alpha$, *i.e.*, $\varepsilon = \|x\|/\alpha$. Since $\|x\| = \|x - 0\| \leq \alpha\varepsilon$, we conclude from (10.19) that $(x, 0) \in R_\varepsilon$. As R_ε is a simulation relation from $S_\tau(\Sigma)$ to $S_\tau(\Sigma)$, for any $\tau \in \mathbb{R}^+$ and any transition $x \xrightarrow[\tau]{\tau} x'$, there exists a transition $0 \xrightarrow[\tau]{\tau} x''$ such that $(x', x'') \in R_\varepsilon$. In other words, for any $t \in \mathbb{R}^+$ we have $(\xi_x(t), \xi_0(t)) \in R_\varepsilon$. If we now denote by ζ the solution of the linear dynamical system $(\mathbb{R}^n, A)$, it is not difficult to see, given (10.2), that $\|\xi_x(t) - \xi_0(t)\| = \|\zeta_x(t) - \zeta_0(t)\|$. Moreover, as $\zeta_0(t) = 0$ for all $t \in \mathbb{R}^+$, and invoking the definition of approximate bisimulation relation, we obtain:

$$\|\zeta_x(t)\| = \|\zeta_x(t) - \zeta_0(t)\| = \|\xi_x(t) - \xi_0(t)\| \leq \varepsilon = \frac{1}{\alpha}\|x\|e^{-0t}.$$

According to Definition 10.1, with $\kappa = 1/\alpha$ and $\lambda = 0$, $x_e = 0$ is a stable equilibrium point for $(\mathbb{R}^n, A)$ and it follows from Theorem 10.4 the existence of a weak Lyapunov function for $(\mathbb{R}^n, A)$. □

Although stability of the origin is sufficient to guarantee the existence of an approximate bisimulation between $S_\tau(\Sigma)$ and $S_\tau(\Sigma)$, Theorem 10.8 does require the stronger assumption of asymptotic stability. It is the dissipative nature of Σ, measured by λ, that compensates the approximation errors, measured by η, introduced in the construction of the transitions of $S_{\tau\eta}(\Sigma)$. This is quantified by inequality (10.7) showing that a smaller λ implies a smaller η.

10.4 Advanced topics

In this section we show that Theorem 10.8 also holds for nonlinear differential equations. The results in this section are based on advanced control theoretic concepts.

We make extensive use of comparison functions of class $\mathcal{K}_\infty$ and $\mathcal{KL}$ to simplify the arguments. A continuous function $\gamma : \mathbb{R}_0^+ \to \mathbb{R}_0^+$ is said to belong to class $\mathcal{K}_\infty$ if it is strictly increasing, $\gamma(0) = 0$, and $\gamma(r) \to \infty$ as $r \to \infty$. A continuous function $\beta : \mathbb{R}_0^+ \times \mathbb{R}_0^+ \to \mathbb{R}_0^+$ is said to belong to class $\mathcal{KL}$ if for each fixed s, the map $\beta(r, s)$ belongs to class $\mathcal{K}_\infty$ with respect to r and, for each fixed r, the map $\beta(r, s)$ is decreasing with respect to s and $\beta(r, s) \to 0$ as $s \to \infty$.

In a nonlinear context, we need to replace the asymptotic stability assumption with the stronger assumption of incremental stability. We first recall the notion of globally asymptotically stable equilibrium point in the framework of comparison functions.

Definition 10.12 (Globally asymptotically stable equilibrium point). *The equilibrium point $x_e \in \mathbb{R}^n$ of a dynamical system $\Sigma = (\mathbb{R}^n, f)$ is* globally asymptotically stable (GAS) *if there exists a $\mathcal{KL}$ function β such that for any $t \in \mathbb{R}_0^+$ and any $x \in \mathbb{R}^n$ the following inequality is satisfied:*

$$\|\xi_x(t) - x_e\| \leq \beta(\|x - x_e\|, t).$$

Definition 10.1 can now be seen as the special case where $x_e = 0$ and $\beta(r, s) = \kappa e^{-\lambda s}\|r\|$. The notion of global asymptotic stability compares trajectories of Σ with the special trajectory $\xi_{x_e}(t) = x_e$. If we compare arbitrary trajectories of Σ we are lead to incremental global asymptotic stability.

Definition 10.13 (Incremental global asymptotic stability). *A dynamical system $\Sigma = (\mathbb{R}^n, f)$ is* incrementally globally asymptotically stable *(δ–GAS) if it is forward complete and there exists a $\mathcal{KL}$ function β such that for any $t \in \mathbb{R}_0^+$ and any $x, x' \in \mathbb{R}^n$ the following inequality is satisfied:*

$$\|\xi_x(t) - \xi_{x'}(t)\| \leq \beta(\|x - x'\|, t). \tag{10.20}$$

In the linear case, δ-GAS degenerates into GAS since $\|\xi_x(t) - \xi_{x'}(t)\| = \|e^{At}x - e^{At}x'\| \leq \|e^{At}\|\|x - x'\| \leq \kappa e^{-\lambda t}\|x - x'\| = \beta(\|x - x'\|, t)$, as required by (10.20). For nonlinear dynamical systems with an equilibrium point x_e, δ-GAS implies GAS. This can be seen by replacing x' and $\xi_{x'}(t)$ with x_e in (10.20). However, the converse is not true: the existence of a globally asymptotically stable equilibrium point does not imply δ-GAS. Nevertheless, δ-GAS can still be characterized by a dissipation inequality.

Definition 10.14 (δ–GAS Lyapunov function). *Let $\Sigma = (\mathbb{R}^n, f)$ be a dynamical system and consider a smooth function $V : \mathbb{R}^n \times \mathbb{R}^n \to \mathbb{R}_0^+$. The function V is a δ–*GAS Lyapunov function *for Σ, if there exist $\mathcal{K}_\infty$ functions $\underline{\alpha}$, $\overline{\alpha}$, and $\lambda \in \mathbb{R}^+$ such that the following inequalities hold for all $x, x' \in \mathbb{R}^n$:*

$$\begin{aligned} \underline{\alpha}(\|x - x'\|) \leq V(x, x') \leq \overline{\alpha}(\|x - x'\|) \\ \frac{\partial V}{\partial x} f(x) + \frac{\partial V}{\partial x'} f(x') \leq -\lambda V(x, x'). \end{aligned}$$

The usual definition of δ-GAS Lyapunov function requires $\frac{\partial V}{\partial x} f(x) + \frac{\partial V}{\partial x'} f(x')$ to be bounded by $-\rho(\|x-x'\|)$ for some $\rho \in \mathcal{K}_\infty$. By modifying V, if necessary, it can be shown that both definitions are in fact equivalent. The following result provides a Lyapunov-like characterization of δ–GAS dynamical systems.

Theorem 10.15. *A dynamical system Σ is δ–GAS if and only if it admits a δ–GAS Lyapunov function.*

In practice, the δ–GAS assumption is tested by searching for a δ–GAS Lyapunov function V. Once V is found, δ-GAS is proved and V can be used to construct an approximately bisimilar symbolic model, as described in the next result.

Theorem 10.16. *Let $\Sigma = (\mathbb{R}^n, f)$ be a δ-GAS dynamical system admitting a δ-GAS Lyapunov function satisfying:*

$$V(x, x') - V(x, x'') \leq \gamma(\|x' - x''\|) \tag{10.21}$$

for some class $\mathcal{K}_\infty$ function γ and for every $x, x', x'' \in \mathbb{R}^n$. For any desired precision $\varepsilon \in \mathbb{R}^+$, for any desired time quantization $\tau \in \mathbb{R}^+$, and for any space quantization $\eta \in \mathbb{R}^+$ satisfying:

$$\eta \leq \min\left\{\gamma^{-1}\left((1 - e^{\lambda\tau})\overline{\alpha}(\varepsilon)\right), \overline{\alpha}^{-1} \circ \underline{\alpha}(\varepsilon)\right\}, \tag{10.22}$$

the relation $R_\varepsilon \subseteq X_\tau \times X_{\tau\eta}$ defined by:

$$R_\varepsilon = \{(x_\tau, x_{\tau\eta}) \in X_\tau \times X_{\tau\eta} \mid V(x_\tau, x_{\tau\eta}) \leq \underline{\alpha}(\varepsilon)\} \tag{10.23}$$

is an ε-approximate bisimulation relation between $S_\tau(\Sigma)$ and $S_{\tau\eta}(\Sigma)$.

Proof. We only provide a sketch since the proof proceeds along the same lines of the proof of Theorem 10.8. The first two requirements in Definition 9.2 follow by construction of R_ε and the third requirement requires a simple modification of the sequence of inequalities (10.11) through (10.15):

$$\begin{aligned}
V(x'_\tau, x'_{\tau\eta}) &\leq V(x'_\tau, x''_{\tau\eta}) + \gamma\left(\|x'_{\tau\eta} - x''_{\tau\eta}\|\right) \\
&\leq V(\xi_{x_\tau}(\tau), \xi_{x_{\tau\eta}}(\tau)) + \gamma(\eta) \\
&\leq e^{-\lambda\tau} V(\xi_{x_\tau}(0), \xi_{x_{\tau\eta}}(0)) + \gamma(\eta) \\
&\leq e^{-\lambda\tau}\underline{\alpha}(\varepsilon) + \gamma(\eta) \\
&\leq \underline{\alpha}(\varepsilon)
\end{aligned}$$

where the first inequality is now a consequence of (10.21). □

The reader can easily verify that Theorem 10.8 can be recovered as a special case of the preceding result by taking $\overline{\alpha}(r) = \overline{\alpha}r$, $\underline{\alpha}(r) = \underline{\alpha}r$, and $\gamma(r) = \gamma r$.

Inequality (10.21) is no longer guaranteed to hold, as was the case for linear dynamical systems, and is thus an additional assumption. Nevertheless, (10.21) is satisfied provided that V is continuously differentiable and we work on a convex compact set.

10.5 Notes

The study of abstractions based on approximate similarity is very recent. This chapter is based on [PGT08, GPT09] where it is shown that, under suitable stability assumptions, nonlinear control systems and nonlinear switched systems admit finite-state approximate bisimilar models. While the results in [PGT08] do not require Lyapunov functions, they impose a lower bound on the sampling time τ used to construct $S_{\tau}(\Sigma)$ and $S_{\tau\eta}(\Sigma)$. The approach followed in this chapter originally appeared in [GPT09] and does not impose any constraint on τ. This means that for any desired value of $\tau \in \mathbb{R}^{+}$, and for any desired precision of $\varepsilon \in \mathbb{R}^{+}$, there always exists a space quantization $\eta \in \mathbb{R}^{+}$ satisfying (10.7) for linear dynamical systems, (10.17) for affine dynamical systems, or (10.22) for nonlinear dynamical systems.

The converse result described by Proposition 10.11 appeared in [Tab08a] where the implication (10.19) was taken as an integral part of the definition of approximate simulation. This result shows that Lyapunov functions, a typical weapon in the control theorist arsenal, are essential ingredients in the study of the recently introduced notion of approximate bisimulation for dynamical and control systems. Lyapunov functions also appear, under the name of bisimulation functions, in the study of metrics for dynamical and control systems [GP07], following previous work on probabilistic systems [DGJP99], finite-state systems [dAFS04], and timed automata [HMP05].

Earlier attempts to use incremental stability to simplify verification problems appeared in [GGM06, GP06, FGP06, Gir07] and were preceded by work on approximate verification that did not rely on incremental stability [PVB95, CK01].

The treatment of δ-GAS follows [Ang02] where the proof of Theorem 10.15 can be found. The proofs of Theorem 10.2 and Theorem 10.4 can be found on any book on linear system theory such as [AM97].

Although Theorem 10.8 can potentially be used to construct abstractions of hybrid dynamical systems, no systematic construction procedure appeared in the literature so far. The difficulty lies in identifying easily checkable conditions determining when the entrance of a trajectory in a guard set implies that all the surrounding trajectories also enter the same guard set.

11

Approximate symbolic models for control

This chapter continues the generalization from exact to approximate similarity, now in the context of symbolic models for control. We discuss approximate feedback composition and refinement, and show how the techniques developed in Chapter 10 can be suitably extended to control systems and switched affine systems. Nonlinear extensions of these results are presented as special topics.

Notation

The following notation is used in this chapter. For any matrix $P \in \mathbb{R}^{n\times n}$, P^T denotes the transposed matrix. Matrix P is said to be symmetric if $P^T = P$, and is said to be positive definite if for every $x \in \mathbb{R}^n$, $x \neq 0$ implies $x^T P x > 0$. We denote by $\mathcal{SP}(n)$ the set of all symmetric and positive definite matrices in $\mathbb{R}^{n\times n}$. The minimum and the maximum eigenvalues of a matrix $P \in \mathbb{R}^{n\times n}$ are denoted by $\lambda_m(P)$ and $\lambda_M(P)$, respectively. For any $x \in \mathbb{R}^n$, $\|x\|$ represents the Euclidean norm of x defined by $\|x\| = \left(x_1^2 + x_2^2 + \ldots + x_n^2\right)^{\frac{1}{2}}$ where x_i is the ith component of the vector x. This norm induces a norm in the space of matrices that can be computed as $\|A\| = \lambda_M^{\frac{1}{2}}(A^T A)$ for any $A \in \mathbb{R}^{n\times m}$. The exponential of any matrix $A \in \mathbb{R}^{n\times n}$ is denoted by e^A and is the analytic function $\sum_{i=0}^{\infty} \frac{1}{i!}A^i$. The ball of radius $r \in \mathbb{R}_0^+$ centered at $x \in \mathbb{R}^n$ is denoted by $\mathcal{B}_r(x)$ and defined as the set of all the points $x' \in \mathbb{R}^n$ satisfying $\|x - x'\| \leq r$. If $Z \subseteq \mathbb{R}^n$ and $\eta \in \mathbb{R}^+$, $[Z]_\eta$ denotes the subset $[Z]_\eta \subseteq Z$ defined by:

$$[Z]_\eta = \left\{ z \in Z \mid z_i = k_i \frac{2}{\sqrt{n}}\eta \text{ for some } k_i \in \mathbb{Z} \text{ and } i = 1, 2, \ldots, n \right\}.$$

Note that we can cover Z by balls of radius η centered at the points in $[Z]_\eta$. This observation is used several times in this chapter.

Given a subset $W \subseteq Z$ we denote by $\imath : W \hookrightarrow Z$ the natural inclusion of W in Z taking $w \in W$ to $\imath(w) = w \in Z$. The identity map on Z is denoted by $1_Z : Z \to Z$ while $\pi_X : X_a \times X_b \times U_a \times U_b \to X_a \times X_b$ denotes the projection

P. Tabuada, *Verification and Control of Hybrid Systems: A Symbolic Approach*,
DOI: 10.1007/978-1-4419-0224-5_11,

sending $(x_a, x_b, u_a, u_b) \in X_a \times X_b \times U_a \times U_b$ to $(x_a, x_b) \in X_a \times X_b$. A relation $R \subseteq Z \times W$ is surjective when for every $w \in W$ there is a $z \in Z$ satisfying $(z, w) \in R$.

A metric on a set Z is a function $\mathbf{d} : Z \times Z \to \mathbb{R}_0^+$ satisfying: $\mathbf{d}(z, z') = 0$ iff $z = z'$; $\mathbf{d}(z, z') + \mathbf{d}(z', z'') \geq \mathbf{d}(z, z'')$; $\mathbf{d}(z, z') = \mathbf{d}(z', z)$. A metric $\mathbf{d}$ is said to be norm-induced if $\mathbf{d}(x, y) = \|x - y\|$ for some norm $\|\cdot\|$ and for every $x, y \in Z$. A metric $\mathbf{d} : Z \times Z \to \mathbb{R}_0^+$ on the set Z induces a pseudo-metric on 2^Z, the set of all subsets of Z. Such pseudo-metric, called the Hausdorff pseudo-metric and denoted by $\mathbf{d}_h$, is defined by $\mathbf{d}_h(K, W) = \max\left\{\overrightarrow{\mathbf{d}_h}(K, W), \overrightarrow{\mathbf{d}_h}(W, K)\right\}$, where $\overrightarrow{\mathbf{d}_h}(K, W) = \sup_{k \in K} \inf_{w \in W} \mathbf{d}(k, w)$ is the directed Hausdorff pseudo-metric and $K, W \subseteq Z$. We recall that the Hausdorff pseudo-metric $\mathbf{d}_h$ satisfies all the requirements of a metric except that $W = W'$ implies $\mathbf{d}_h(W, W') = 0$ but $\mathbf{d}_h(W, W') = 0$ does not imply $W = W'$.

A function $f :]a, b[\to \mathbb{R}^n$, $a, b \in \mathbb{R}$, is said to be piecewise continuous if there exists an ordered sequence of real numbers $a = i_1 < i_2 < \ldots < i_k = b$ such that for every $j \in \{1, 2, \ldots, k-1\}$, the restriction of f to the interval $]i_j, i_{j+1}[$ is continuous. A piecewise continuous function $f :]a, b[\to \mathbb{R}^n$ is essentially bounded if there exists a compact set $K \subset \mathbb{R}^n$ such that $f(t) \in K$ for almost all $t \in]a, b[$. When $f :]a, b[\to \mathbb{R}^n$ is an essentially bounded piecewise continuous function, the supremum norm of f, denoted by $\|f\|$, is the supremum of the set $\{r \in \mathbb{R}_0^+ \mid \exists\, t \in]a, b[\quad r = \|f(t)\| \wedge f(t) \in K\}$. The domain of a function $f : Z \to W$ is denoted by $\operatorname{dom} f$.

11.1 Stability of linear control systems

We review a few stability results needed for the study of approximate simulations and bisimulations. The reader is expected to have read Section 8.1.2 where several concepts related to control systems were introduced. Here, we consider affine control systems described by the affine differential equation:

$$\frac{d}{dt}\xi = A\xi + C\chi + D\delta + h \tag{11.1}$$

with $\xi(t) \in \mathbb{R}^n$, $\chi(t) \in \mathbb{R}^m$, $\delta(t) \in \mathbb{R}^l$, $\chi \in \mathcal{C}$, $\delta \in \mathcal{D}$, $A \in \mathbb{R}^{n \times n}$, $C \in \mathbb{R}^{n \times m}$, $D \in \mathbb{R}^{n \times l}$, $h \in \mathbb{R}^n$, and $t \in \mathbb{R}_0^+$. We distinguish between two different kinds of inputs: control inputs χ, and disturbance inputs δ. We are thus taking $\mathcal{U} = \mathcal{C} \times \mathcal{D}$ and $\upsilon = (\chi, \delta)$ according to the notion of continuous-time control system introduced in Section 8.1.2. Independently of the nature of the inputs, a solution to (11.1) can always be written in the form:

$$\xi_{x\chi\delta}(\tau) = e^{A\tau}x + \int_0^\tau e^{A(\tau - t)}\left(C\chi(t) + D\delta(t) + h\right) dt. \tag{11.2}$$

Affine control systems are denoted by the septuple $\Sigma = (\mathbb{R}^n, \mathcal{C}, \mathcal{D}, A, C, D, h)$ or by the sextuple $\Sigma = (\mathbb{R}^n, \mathcal{C}, \mathcal{D}, A, C, D)$ when $h = 0$. In the later case we speak of a linear control system. Although we are interested in the slightly more general class of affine control systems, it is sufficient to consider the stability properties of linear control systems. In some of the results we will assume the absence of disturbances, *i.e.*, $D = 0$. In such cases we denote Σ by the quadruple $(\mathbb{R}^n, \mathcal{C}, A, C)$.

Definition 11.1 (Input-to-state stability). *A linear control system* $(\mathbb{R}^n, \mathcal{C}, \mathcal{D}, A, C, D)$ *is said to be* input-to-state stable (ISS) *when there exist constants* $\kappa, \lambda, \rho_c, \rho_d \in \mathbb{R}^+$ *such that for any* $x \in \mathbb{R}^n$*, any* $\chi \in \mathcal{C}$*, any* $\delta \in \mathcal{D}$*, and any* $t \in \mathbb{R}^+$*, the following inequality is satisfied:*

$$\|\xi_{x\chi\delta}(t)\| \leq \kappa e^{-\lambda t}\|x\| + \rho_c\|\chi\| + \rho_d\|\delta\|. \tag{11.3}$$

Inequality (11.3) extends inequality (10.3) from linear dynamical systems to linear control systems. The next step is to extended also the concept of Lyapunov function.

Definition 11.2 (ISS Lyapunov function). *Let* $(\mathbb{R}^n, \mathcal{C}, \mathcal{D}, A, C, D)$ *be a linear control system and consider a function* $V : \mathbb{R}^n \to \mathbb{R}$ *satisfying the following three properties:*

1. *V is continuous on $\mathbb{R}^n$ and smooth on $\mathbb{R}^n \backslash \{0\}$;*
2. *$V(x) \geq 0$ for all $x \in \mathbb{R}^n$;*
3. *$V(x) = 0$ implies $x = 0$.*

The function V is an ISS-Lyapunov function *for Σ if there exist constants $\lambda, \sigma_c, \sigma_d \in \mathbb{R}^+$ such that for all $x \in \mathbb{R}^n \backslash \{0\}$, $c \in \mathbb{R}^m$, and $d \in \mathbb{R}^l$, the following inequality holds:*

$$\frac{\partial V}{\partial x}(Ax + Cc + Dd) \leq -\lambda V(x) + \sigma_c\|c\| + \sigma_d\|d\|. \tag{11.4}$$

Inequality (11.4) entails the differential inequality:

$$\frac{d}{dt}V \circ \xi \leq -\lambda V \circ \xi + \sigma_c\|\chi\| + \sigma_d\|\delta\|$$

that can be integrated to provide the estimate:

$$\begin{aligned} V \circ \xi(t) &\leq e^{-\lambda t}V(\xi(0)) + \frac{\sigma_c\|\chi\|}{\lambda}(1 - e^{-\lambda t}) + \frac{\sigma_d\|\delta\|}{\lambda}(1 - e^{-\lambda t}) \\ &\leq e^{-\lambda t}V(\xi(0)) + \frac{\sigma_c}{\lambda}\|\chi\| + \frac{\sigma_d}{\lambda}\|\delta\|. \end{aligned} \tag{11.5}$$

Inequality (11.5) can be combined with (10.5) to fully characterize ISS in terms of ISS-Lyapunov functions as stated in the next result.

Theorem 11.3. *A linear control system Σ is input-to-state stable iff Σ admits an ISS-Lyapunov function.*

For linear systems, the above theorem can be strengthened by asserting that ISS implies the existence of an ISS-Lyapunov function of the form $V(x) = \sqrt{x^T P x}$ with $P \in \mathcal{SP}(n)$. Moreover, it can be shown that a linear control system $(\mathbb{R}^n, \mathcal{C}, \mathcal{D}, A, C, D)$ is ISS iff the origin is an asymptotically stable equilibrium point for the linear dynamical system $(\mathbb{R}^n, A)$. The ISS assumption is thus very simple to check since Theorem 10.2 asserts that it suffices to determine if all the eigenvalues of the matrix A have negative real part.

Although input-to-state stability is the assumption upon which all the results in this chapter rely, there is a straightforward extension to a wider class of control systems. When a linear control system Σ is not ISS, it may be rendered ISS by suitably designing a linear feedback control law $\chi = K\xi + \chi'$ transforming Σ into the linear control system defined by:

$$\frac{d}{dt}\xi = (A + CK)\xi + C\chi' + D\delta$$

with new control input χ'. ISS is achieved whenever K makes the real part of the eigenvalues of $A + CK$ negative. The results in this chapter remain valid for this larger class of systems even though, for simplicity, we will directly assume input-to-state stability.

11.2 Control and switched systems as systems

11.2.1 Control Systems

In Chapter 10 we introduced the system $S_\tau(\Sigma)$ describing the time-triggered sampled version of a given dynamical system Σ. A simple generalization is available for control systems.

Definition 11.4. *The system $S_\tau = (X_\tau, U_\tau, \underset{\tau}{\longrightarrow})$ associated with a control system $\Sigma = (\mathbb{R}^n, \mathcal{C} \times \mathcal{D}, f)$ and with $\tau \in \mathbb{R}^+$ consists of:*

- $X_\tau = \mathbb{R}^n$;
- $U_\tau = \{\chi \in \mathcal{C} \mid \operatorname{dom}\chi = [0, \tau]\}$;
- $x \overset{\chi}{\underset{\tau}{\longrightarrow}} x'$ *if there exist* $\chi \in U_\tau$, $\delta \in \mathcal{D}$, *and a trajectory* $\xi_{x\chi\delta} : [0, \tau] \to \mathbb{R}^n$ *of* Σ *satisfying* $\xi_{x\chi\delta}(\tau) = x'$;
- $Y_\tau = \mathbb{R}^n$;
- $H_\tau = \imath : X_\tau \hookrightarrow \mathbb{R}^n$.

The output set $Y_\tau = \mathbb{R}^n$ of $S_\tau(\Sigma)$ is naturally equipped with the norm-induced metric $\mathbf{d}(y, y') = \|y - y'\|$. In addition to control systems, we also consider switched systems.

11.2.2 Switched systems

Switched systems are a class of hybrid dynamical systems frequently arising in embedded control applications. We restrict the discussion to the case where the continuous-time dynamics in each finite state is given by an affine dynamical system.

Definition 11.5 (Switched affine system). *A hybrid dynamical system:*

$$\Sigma = (S_a, \{\mathrm{In}_{x_a}\}_{x_a \in X_a}, \{\mathrm{Gu}_{t_a}\}_{t_a \in \underset{a}{\longrightarrow}}, \{\mathrm{Re}_{t_a}\}_{t_a \in \underset{a}{\longrightarrow}}, \{f_{x_a}\}_{x_a \in X_a})$$

is said to be a switched affine system *if the following conditions are satisfied:*

1. $U_a = X_a$;
2. $\underset{a}{\longrightarrow} = \{(x_a, u_a, x'_a) \in X_a \times X_a \times X_a \mid u_a = x'_a\}$;
3. $\mathrm{In}_{x_a} = \mathbb{R}^n$ *for every* $x_a \in X_a$;
4. $\mathrm{Gu}_{(x_a,u_a,x'_a)} = \mathbb{R}^n$ *for every* $(x_a, u_a, x'_a) \in \underset{a}{\longrightarrow}$;
5. $\mathrm{Re}_{(x_a,u_a,x'_a)}(x_b) = x_b$ *for every* $(x_a, u_a, x'_a) \in \underset{a}{\longrightarrow}$ *and* $x_b \in \mathrm{In}_{x_a}$;
6. $f_{x_a}(x_b) = A_{x_a} x_b + h_{x_a}$ *for some matrix* $A_{x_a} \in \mathbb{R}^{n \times n}$*, some vector* $h_{x_a} \in \mathbb{R}^n$*, and all* $x_a \in X_a$*,* $x_b \in \mathrm{In}_{x_a}$.

In a switched affine system it is possible, at any time and independently of the infinite state, to switch from any finite state to any other finite state without changing the infinte part of the state. This possibility is described by the several requirements in Definition 11.5. The first two requirements ask that for every two finite states $x_a, x'_a \in X_a$ there exists one and only one transition between them: $x_a \overset{x'_a}{\underset{a}{\longrightarrow}} x'_a$. The third and sixth requirements ask that in each finite state $x_a \in X_a$, the switched system behaves like the affine dynamical system $(\mathbb{R}^n, A_{x_a}, h_{x_a})$. The fourth requirement allows for discrete transitions to take place at any time and for any value of the infinite part of the state. Finally, the fifth condition declares that discrete transitions do not alter the infinite part of the state. These restrictions also imply that a switched affine system is completely defined by the finite set of states X_a, and the collection of affine dynamical systems $\{(\mathbb{R}^n, A_{x_a}, h_{x_a})\}_{x_a \in X_a}$. For this reason, we also denote a switched affine system by the triple $\Sigma = (X_a, \mathbb{R}^n, \{A_{x_a}, h_{x_a}\}_{x_a \in X_a})$.

Example 11.6. Switched affine systems provide a useful framework for switching control. Suppose that several affine controllers:

$$c = K_1 x + h_1, c = K_2 x + h_2, \ldots, c = K_p x + h_p,$$

have been designed to control the linear system:

$$\dot{\xi} = A\xi + C\chi, \qquad \xi(t) \in \mathbb{R}^n, \chi(t) \in \mathbb{R}^m, t \in \mathbb{R}^+_0.$$

If these controllers can be used independently of the infinite state $x \in \mathbb{R}^n$, we have a switched affine system Σ described by:

$$(\{1, 2, \ldots, p\}, \mathbb{R}^n, \{A + CK_i, Ch_i\}_{i \in \{1,2,\ldots,p\}}).$$

A software module deciding which controller is executed and when, can now be seen as a supervisory controller acting on the switched affine system Σ. $\triangleleft$

When switched affine systems are viewed as models for switching control, the supervisory controller is typically implemented as a periodic task, with period τ, running on a microprocessor. This implies that discrete transitions only happen at instants that are integer multiples of τ. An appropriate model for this kind of system is $S_\tau(\Sigma)$, capturing only transitions of duration τ.

Definition 11.7. *The system* $S_\tau(\Sigma) = (X_\tau, U_\tau, \underset{\tau}{\longrightarrow}, Y_\tau, H_\tau)$ *associated with a switched affine system* $\Sigma = (X_a, \mathbb{R}^n, \{A_{x_a}, h_{x_a}\}_{x_a \in X_a})$ *and with* $\tau \in \mathbb{R}^+$ *consists of:*

- $X_\tau = \mathbb{R}^n$;
- $U_\tau = X_a$;
- $x \xrightarrow[\tau]{u_a} x'$ *if there exists a solution* $\xi_x : [0, \tau] \to \mathbb{R}^n$ *of the affine dynamical system* $(\mathbb{R}^n, A_{u_a}, h_{u_a})$ *satisfying* $\xi_x(\tau) = x'$;
- $Y_\tau = \mathbb{R}^n$;
- $H_\tau = \imath : X_\tau \hookrightarrow \mathbb{R}^n$.

Note that $S_\tau(\Sigma)$ is both infinite-state as well as metric with a norm-induced metric.

11.3 Approximate feedback composition and controller refinement

The controller refinement process carries over, mutatis mutandis, from the exact to the approximate case. We recall that in Chapter 1 we simplified the representation of the composition $S_a \times_{\mathcal{I}} S_b$ whenever the interconnection relation $\mathcal{I}$ satisfied the condition:

$$(x_a, x_b) \in \pi_X(\mathcal{I}) \implies H_a(x_a) = H_b(x_b).$$

In the current approximate context, we consider the generalized condition:

$$(x_a, x_b) \in \pi_X(\mathcal{I}) \implies \mathbf{d}(H_a(x_a), H_b(x_b)) \leq \varepsilon$$

and make the additional assumption that $\mathbf{d}$ is norm-induced. Note that this assumption entails that $Y_a = Y_b$ are normed vector spaces with the same norm. Under these assumptions, we denote the composition by:

$$S_a \times_{\mathcal{I}}^{\varepsilon} S_b = (X_{ab}, X_{ab0}, U_{ab}, \xrightarrow[ab]{}, Y_{ab}, H_{ab})$$

and simplify its representation to:

- $X_{ab} = \pi_X(\mathcal{I})$;
- $X_{ab0} = X_{ab} \cap (X_{a0} \times X_{b0})$;
- $U_{ab} = U_a \times U_b$;
- $(x_a, x_b) \xrightarrow[ab]{u_a, u_b} (x'_a, x'_b)$ if the following three conditions hold:
  1. $x_a \xrightarrow[a]{u_a} x'_a$ in S_a;
  2. $x_b \xrightarrow[b]{u_b} x'_b$ in S_b;
  3. $(x_a, x_b, u_a, u_b) \in \mathcal{I}$;
- $Y_{ab} = Y_a = Y_b$;
- $H_{ab}(x_a, x_b) = \frac{1}{2}(H_a(x_a) + H_b(x_b))$.

The apparently arbitrary choice of output map is justified by the following three important properties of approximate composition:

1. $S_a \times_{\mathcal{I}}^{\varepsilon} S_b$ is commutative, *i.e.*, $S_a \times_{\mathcal{I}}^{\varepsilon} S_b \cong_{\mathcal{S}} S_b \times_{\mathcal{I}}^{\varepsilon} S_a$;
2. $S_a \times_{\mathcal{I}}^{\varepsilon} S_b$ generalizes exact composition, *i.e.*, $S_a \times_{\mathcal{I}}^{0} S_b = S_a \times_{\mathcal{I}} S_b$;
3. $S_a \times_{\mathcal{I}}^{\varepsilon} S_b$ satisfies the following version of Proposition 6.3.

Proposition 11.8. *Let S_a and S_b be metric systems with $Y_a = Y_b$ normed vector spaces with the same norm-induced metric, and let $\mathcal{I}$ be an interconnection relation satisfying:*

$$(x_a, x_b) \in \pi_X(\mathcal{I}) \implies \mathbf{d}(H_a(x_a), H_b(x_b)) \leq \varepsilon.$$

Then, the following holds:

- $S_a \times_{\mathcal{I}}^{\varepsilon} S_b \preceq_{\mathcal{S}}^{\frac{1}{2}\varepsilon} S_a$;
- $S_a \times_{\mathcal{I}}^{\varepsilon} S_b \preceq_{\mathcal{S}}^{\frac{1}{2}\varepsilon} S_b$.

Proof. The proof of this result is the same as the proof of its exact counterpart, Proposition 6.3, except for the computation of the precision. We thus focus on this part and consider only $S_a \times_{\mathcal{I}}^{\varepsilon} S_b \preceq_{\mathcal{S}}^{\frac{1}{2}\varepsilon} S_a$ since the case $S_a \times_{\mathcal{I}}^{\varepsilon} S_b \preceq_{\mathcal{S}}^{\frac{1}{2}\varepsilon} S_b$ can be similarly proved. The desired $\frac{1}{2}\varepsilon$-approximate simulation relation from $S_a \times_{\mathcal{I}}^{\varepsilon} S_b$ to S_a is given by:

$$R_a = \{((x_a, x_b), x'_a) \in X_{ab} \times X_a \mid x_a = x'_a\}.$$

For any $((x_a, x_b), x_a) \in R_a$ it is simple to see that:

$$\begin{aligned}\mathbf{d}(H_{ab}(x_a, x_b), H_a(x_a)) &= \left\| \frac{1}{2}H_a(x_a) + \frac{1}{2}H_b(x_b) - H_a(x_a) \right\| \\ &= \left\| -\frac{1}{2}H_a(x_a) + \frac{1}{2}H_b(x_b) \right\| \\ &= \frac{1}{2}\mathbf{d}(H_a(x_a), H_b(x_b)) \leq \frac{1}{2}\varepsilon\end{aligned}$$

since $(x_a, x_b) \in \pi_X(\mathcal{I})$. □

With the notion of approximate composition at our disposal we venture into approximate feedback composition.

Definition 11.9 (Approximate feedback composition). *A system S_c is said to be ε-*approximately feedback composable *with a system S_a, if there exists an ε-approximate alternating simulation relation R from S_c to S_a. When S_c is ε-approximate feedback composable with S_a, the feedback composition of S_c and S_a, with interconnection relation $\mathcal{F} = R^e$, is given by $S_c \times^{\varepsilon}_{\mathcal{F}} S_a$.*

Proposition 8.7 also admits an approximate version.

Proposition 11.10. *Let S_a, S_b, and S_c be systems with the same output set, assume that S_c is ${}_c\varepsilon_a$-approximately feedback composable with S_a, and let ${}_cR_a$ be the corresponding ${}_c\varepsilon_a$-approximate alternating simulation relation. If there exists a ${}_a\varepsilon_b$-approximate alternating simulation relation ${}_aR_b$ from S_a to S_b then $S_c \times^{{}_c\varepsilon_a}_{{}_cR^e_a} S_a$ is feedback composable with S_b and the corresponding $({}_c\varepsilon_a + {}_a\varepsilon_b)$-approximate alternating simulation relation is given by:*

$$ {}_{ca}R_b = \{((x_c, x_a), x_b) \in (X_c \times X_a) \times X_b \mid (x_c, x_a) \in {}_cR_a \wedge (x_a, x_b) \in {}_aR_b\}.$$

The proof of this result consists in inserting the word approximate in several locations along the proof of Proposition 8.7 and is therefore omitted. Proposition 11.10 suggests how to refine a controller S_{cont} synthesized to solve a simulation game for an approximate finite-state abstraction S_{abs} of S and a specification S_{spec}. If the simulation game is solved exactly, *i.e.*, with $\varepsilon = 0$, we have:

$$S_{cont} \times_{\mathcal{F}} S_{abs} \preceq^0_{\mathcal{S}} S_{spec}.$$

Assuming the abstraction S_{abs} to be related to the original system S by an ε-approximate alternating simulation relation, we can invoke Proposition 11.10 to conclude that $S_{cont} \times_{\mathcal{F}} S_{abs}$ is ε-approximately feedback composable with S. Therefore, using $S'_{cont} = S_{cont} \times_{\mathcal{F}} S_{abs}$ as a controller for S we obtain:

$$S'_{cont} \times^{0+\varepsilon}_{\mathcal{G}} S \preceq^{\frac{1}{2}\varepsilon}_{\mathcal{S}} S'_{cont} = S_{cont} \times_{\mathcal{F}} S_{abs} \preceq^0_{\mathcal{S}} S_{spec}$$

which shows the specification approximately simulating the controlled system $S'_{cont} \times^{\varepsilon}_{\mathcal{G}} S$ with precision $\frac{1}{2}\varepsilon$.

11.4 Symbolic models for affine control systems

The abstractions constructed in Chapter 10 for dynamical systems relied on quantizing the set of states and approximating the transitions of $S_\tau(\Sigma)$. A natural generalization to control systems leads to the following construction.

Definition 11.11. *The system* $S_{\tau\eta} = (X_{\tau\eta}, U_{\tau\eta}, \underset{\tau\eta}{\longrightarrow}, Y_{\tau\eta}, H_{\tau\eta})$ *associated with a control system* $\Sigma = (\mathbb{R}^n, \mathcal{C} \times \mathcal{D}, f)$ *and with* $\tau, \eta \in \mathbb{R}^+$ *consists of:*

- $X_{\tau\eta} = [\mathbb{R}^n]_\eta$;
- $U_{\tau\eta} = \{\chi \in \mathcal{C} \mid \operatorname{dom}\chi = [0, \tau]\}$;
- $x \xrightarrow[\tau\eta]{\chi} x'$ *if there exist* $\chi \in U_{\tau\eta}$, $\delta \in \mathcal{D}$, *and a trajectory* $\xi_{x\chi\delta} : [0, \tau] \to \mathbb{R}^n$ *of* Σ *satisfying* $\|\xi_{x\chi\delta}(\tau) - x'\| \leq \eta$;
- $Y_{\tau\eta} = \mathbb{R}^n$;
- $H_{\tau\eta} = \imath : X_{\tau\eta} \hookrightarrow \mathbb{R}^n$.

The system $S_{\tau\eta}(\Sigma)$ can be regarded as a time and space quantization of a control system Σ. It is constructed by approximating the transitions of $S_\tau(\Sigma)$ so as to enforce departure from and arrival at states in $X_{\tau\eta} = [\mathbb{R}^n]_\eta$. This construction is not guaranteed to result in a system approximately simulated by $S_\tau(\Sigma)$ since the mismatch between outputs of $S_{\tau\eta}(\Sigma)$ and $S_\tau(\Sigma)$ can grow without bounds along any two external behaviors. In Chapter 10 we relied on asymptotic stability to overcome this difficulty in the context of dynamical systems. A similar strategy can be employed for control systems in order to establish the existence of an approximate alternating simulation relation from $S_{\tau\eta}(\Sigma)$ to $S_\tau(\Sigma)$. Moreover, such relation would desirably be surjective since this allows us to relate any state of $S_\tau(\Sigma)$ to a state of $S_{\tau\eta}(\Sigma)$ for which a controller can be designed.

Theorem 11.12. *Let* $\Sigma = (\mathbb{R}^n, \mathcal{C}, \mathcal{D}, A, C, D, h)$ *be an affine control system and assume that the linear dynamical system,* $(\mathbb{R}^n, A)$ *admits a Lyapunov function* V *of the form* $V(x) = \sqrt{x^T P x}$ *with* $P \in \mathcal{SP}(n)$. *For any desired precision* $\varepsilon \in \mathbb{R}^+$, *for any desired time quantization* $\tau \in \mathbb{R}^+$, *and for any space quantization* $\eta \in \mathbb{R}^+$ *satisfying:*

$$\eta \leq \min\left\{\gamma^{-1}\underline{\alpha}\varepsilon\left(1 - e^{-\lambda\tau}\right), \overline{\alpha}^{-1}\underline{\alpha}\varepsilon\right\}, \tag{11.6}$$

the relation $R_\varepsilon \subseteq X_{\tau\eta} \times X_\tau$ *defined by:*

$$R_\varepsilon = \{(x_{\tau\eta}, x_\tau) \in X_{\tau\eta} \times X_\tau \mid V(x_\tau - x_{\tau\eta}) \leq \underline{\alpha}\varepsilon\} \tag{11.7}$$

is a surjective ε*-approximate alternating simulation relation from* $S_{\tau\eta}(\Sigma)$ *to* $S_\tau(\Sigma)$.

Proof. The proof consists in showing that R_ε satisfies all the requirements in the definition of approximate alternating simulation relation.

We first note that R_ε is surjective since $\mathbb{R}^n \subseteq \bigcup_{x\in[\mathbb{R}^n]_\eta}\mathcal{B}_\eta(x)$ implies that for every $x_\tau \in X_\tau = \mathbb{R}^n$ there exists $x_{\tau\eta} \in X_{\tau\eta}$ satisfying $\|x_\tau - x_{\tau\eta}\| \leq \eta$. It follows from the sequence of inequalities (10.9) that $(x_{\tau\eta}, x_\tau) \in R_\varepsilon$.

The first requirement in Definition 9.6 follows immediately from the definition of $X_{\tau 0}$ and $X_{\tau\eta 0}$, and from the observation that $x_{\tau\eta 0} \in X_{\tau\eta 0} \subset X_{\tau 0}$ implies $(x_{\tau\eta 0}, x_{\tau 0}) \in R_\varepsilon$ for $x_{\tau 0} = x_{\tau\eta 0}$.

The second requirement is a consequence of the definition of R_ε. If $(x_{\tau\eta}, x_\tau) \in R_\varepsilon$, then $V(x_{\tau\eta} - x_\tau) \leq \underline{\alpha}\varepsilon$ which leads, by (10.5), to $\|x_{\tau\eta} - x_\tau\| \leq \varepsilon$.

We now consider the third requirement which requires us to show that $(x_{\tau\eta}, x_\tau) \in R_\varepsilon$ implies:

$$\forall u_{\tau\eta} \in U_{\tau\eta}(x_{\tau\eta}) \quad \exists u_\tau \in U_\tau(x_\tau) \quad \forall x'_\tau \in \mathrm{Post}_{u_\tau}(x_\tau) \quad \exists x'_{\tau\eta} \in \mathrm{Post}_{u_{\tau\eta}}(x_{\tau\eta})$$

with $(x'_{\tau\eta}, x'_\tau) \in R_\varepsilon$. Fix an input $u_{\tau\eta} \in U_{\tau\eta}(x_{\tau\eta})$ and note that it follows from the definition of $U_{\tau\eta}$ that $u_{\tau\eta} \in U_\tau(x_\tau)$. We then choose u_τ to be $u_{\tau\eta}$, *i.e.*, $u_\tau = u_{\tau\eta}$. Let now $x'_\tau \in \mathrm{Post}_{u_\tau}(x_\tau)$. This means that $x'_\tau = \xi_{x_\tau u_\tau \delta}(\tau)$ for some essentially bounded piecewise continuous curve $\delta \in \mathcal{D}$. Consider a state $x'_{\tau\eta} \in \mathrm{Post}_{u_{\tau\eta}}(x_{\tau\eta})$ satisfying $x_{\tau\eta} \xrightarrow[\tau\eta]{u_{\tau\eta},\delta} x'_{\tau\eta}$ in $S_{\tau\eta}(\Sigma)$ and recall that, by definition of $S_{\tau\eta}(\Sigma)$, we have:

$$\|\xi_{x_{\tau\eta}u_{\tau\eta}\delta}(\tau) - x'_{\tau\eta}\| \leq \eta. \tag{11.8}$$

We claim that $(x'_{\tau\eta}, x'_\tau) \in R_\varepsilon$. To prove the claim, consider the sequence of inequalities:

$$\begin{aligned}
V(x'_\tau, x'_{\tau\eta}) &\leq V\left(x'_\tau - \xi_{x_{\tau\eta}u_{\tau\eta}\delta}(\tau)\right) + \gamma\|\xi_{x_{\tau\eta}u_{\tau\eta}\delta}(\tau) - x'_{\tau\eta}\| \\
&\leq V\left(e^{A\tau}x_\tau + \int_0^\tau e^{A(\tau-t)}\left(Cu_{\tau\eta}(t) + D\delta(t) + h\right)dt\right. \\
&\qquad \left. - e^{A\tau}x_{\tau\eta} - \int_0^\tau e^{A(\tau-t)}\left(Cu_{\tau\eta}(t) + D\delta(t) + h\right)dt\right) + \gamma\eta \\
&\leq V\left(e^{A\tau}x_\tau - e^{A\tau}x_{\tau\eta}\right) + \gamma\eta \\
&\leq V\left(\xi_{x_\tau 00}(\tau) - \xi_{x_{\tau\eta}00}(\tau)\right) + \gamma\eta \\
&\leq e^{-\lambda\tau}V\left(\xi_{x_\tau 00}(0) - \xi_{x_{\tau\eta}00}(0)\right) + \gamma\eta \\
&\leq e^{-\lambda\tau}V\left(x_\tau - x_{\tau\eta}\right) + \gamma\eta \\
&\leq e^{-\lambda\tau}\underline{\alpha}\varepsilon + \gamma\eta \\
&\leq \underline{\alpha}\varepsilon
\end{aligned}$$

where the first, second, fifth, seventh, and eight inequalities are a consequence of (10.6), (11.8), (10.16), (11.7), and (11.6), respectively. □

Although we established the existence of a surjective ε-approximate alternating simulation relation from $S_{\tau\eta}(\Sigma)$ to $S_\tau(\Sigma)$, one problem remains unsolved: how do we compute $S_{\tau\eta}(\Sigma)$? We address this problem in two steps. First, we treat the case where disturbance inputs are absent: $D = 0$. By choosing a finite set $\mathcal{C}$ of control inputs curves, it becomes possible to compute $S_{\tau\eta}(\Sigma)$ using numerical methods. The errors introduced by numerical simulation can be explicitly accounted for, as discussed in Chapter 10. In practice, the choice of the set $\mathcal{C}$ is based on domain knowledge about the system and problem being solved. When a solution to a control synthesis problem fails to exist for the abstraction, one can choose a larger set $\mathcal{C}$ and compute a new and more faithful abstraction of the system to be controlled. Ideally, one would like to avoid this iterative process and construct directly a symbolic model that can be used to prove or disprove the existence of a controller. This is possible for the important case where the inputs are kept constant during the intervals $[0, \tau]$, commonly referred to as digital control or sampled-data control. The appropriate system model for this situation is the abstraction $S_{\tau\eta\omega}(\Sigma)$.

Definition 11.13. *The system* $S_{\tau\eta\omega} = (X_{\tau\eta\omega}, U_{\tau\eta\omega}, \underset{\tau\eta\omega}{\longrightarrow}, Y_{\tau\eta\omega}, H_{\tau\eta\omega})$ *associated with a control system* $\Sigma = (\mathbb{R}^n, \mathcal{C} \times \mathcal{D}, f)$ *and with* $\tau, \eta, \omega \in \mathbb{R}^+$ *consists of:*

- $X_{\tau\eta\omega} = [\mathbb{R}^n]_\eta$;
- $U_{\tau\eta\omega} = \{\chi \in \mathcal{C} \mid \chi(t) = \chi(t') \in [\mathbb{R}^m]_\omega \quad \forall t, t' \in [0, \tau] = \operatorname{dom} \chi\}$;
- $x \underset{\tau\eta\omega}{\overset{\chi}{\longrightarrow}} x'$ *if there exist* $\chi \in U_{\tau\eta\omega}$, $\delta \in \mathcal{D}$, *and a trajectory* $\xi_{x\chi\delta} : [0, \tau] \to \mathbb{R}^n$ *of* Σ *satisfying* $\|\xi_{x\chi\delta}(\tau) - x'\| \leq \eta$;
- $Y_{\tau\eta\omega} = \mathbb{R}^n$;
- $H_{\tau\eta\omega} = \imath : X_{\tau\eta\omega} \hookrightarrow \mathbb{R}^n$.

The assumption of piecewise constant inputs is satisfied by most embedded control systems implemented in digital platforms. The frequency of the updates is dictated by the dynamics of the physical system being controlled and by the frequency of the embedded microprocessor executing the control software. Under this assumption we can strengthen Theorem 11.12 from simulation to bisimulation.

Theorem 11.14. *Let* $\Sigma = (\mathbb{R}^n, \mathcal{C}, A, C, h)$ *be an affine control system where* $\mathcal{C}$ *is the set of all constant curves, and assume that the linear control system* $(\mathbb{R}^n, \mathcal{C}, A, C)$ *admits an ISS-Lyapunov function* V *of the form* $V(x) = \sqrt{x^T P x}$ *with* $P \in \mathcal{SP}(n)$. *For any desired precision* $\varepsilon \in \mathbb{R}^+$, *for any desired time quantization* $\tau \in \mathbb{R}^+$, *for any desired input quantization* $\omega \in \mathbb{R}^+$, *and for any space quantization* $\eta \in \mathbb{R}^+$ *satisfying:*

$$\eta \leq \min\left\{\gamma^{-1}\underline{\alpha}\varepsilon\left(1 - e^{-\lambda\tau}\right) - \gamma^{-1}\lambda^{-1}\sigma_c\,\omega, \overline{\alpha}^{-1}\underline{\alpha}\varepsilon\right\}, \tag{11.9}$$

the relation $R_\varepsilon \subseteq X_{\tau\eta} \times X_\tau$ *defined by:*

$$R_\varepsilon = \{(x_{\tau\eta}, x_\tau) \in X_{\tau\eta} \times X_\tau \mid V(x_\tau - x_{\tau\eta}) \leq \underline{\alpha}\varepsilon\} \tag{11.10}$$

is an ε*-approximate bisimulation relation between* $S_\tau(\Sigma)$ *and* $S_{\tau\eta}(\Sigma)$.

Inequality (11.9) describes the tradeoff between precision, time quantization, space quantization, and input quantization. It specializes to (10.7), when inputs are absent, thus making Theorem 10.8 a special case of Theorem 11.14.

Proof. We only present the main steps since the proof mirrors the proof of Theorem 11.12. The first important step is to show that the third requirement in Definition 9.2 holds. For this, we consider a pair $(x_\tau, x_{\tau\eta\omega}) \in R_\varepsilon$, we assume that $x_\tau \xrightarrow[\tau]{u_\tau} x'_\tau$, and we seek to show that $(x'_\tau, x'_{\tau\eta\omega}) \in R_\varepsilon$ where $x'_{\tau\eta\omega}$ satisfies $x_{\tau\eta\omega} \xrightarrow[\tau\eta\omega]{u_{\tau\eta\omega}} x'_{\tau\eta\omega}$ for an input $u_{\tau\eta\omega} \in U_{\tau\eta\omega}(x_{\tau\eta\omega})$ close to u_τ in the sense:

$$\|u_{\tau\eta\omega} - u_\tau\| \leq \omega.$$

Note that such input always exists since $\mathbb{R}^m \subseteq \bigcup_{u \in [\mathbb{R}^m]_\omega} \mathcal{B}_\omega(u)$. The membership $(x'_\tau, x'_{\tau\eta\omega}) \in R_\varepsilon$ follows from the following sequence of inequalities where we use $x'' = x_\tau - x_{\tau\eta\omega}$, $u'' = u_\tau - u_{\tau\eta\omega}$, and the inequality (11.5):

$$V(x'_\tau - x'_{\tau\eta\omega}) \leq V(x'_\tau - \xi_{x_{\tau\eta\omega}u_{\tau\eta\omega}}(\tau)) + \gamma\|\xi_{x_{\tau\eta\omega}u_{\tau\eta\omega}}(\tau) - x'_{\tau\eta\omega}\| \quad (11.11)$$

$$\leq V\left(e^{A\tau}x_\tau + \int_0^\tau e^{A(\tau-t)}(Cu_\tau + h)dt \right. \quad (11.12)$$

$$\left. -e^{A\tau}x_{\tau\eta\omega} - \int_0^\tau e^{A(\tau-t)}(Cu_{\tau\eta\omega} + h)dt\right) + \gamma\eta \quad (11.13)$$

$$\leq V\left(e^{A\tau}x'' + \int_0^\tau e^{A(\tau-t)}Cu''dt\right) + \gamma\eta \quad (11.14)$$

$$\leq V \circ \xi_{x''u''}(\tau) + \gamma\eta \quad (11.15)$$

$$\leq e^{-\lambda\tau}V \circ \xi_{x''u''}(0) + \frac{\sigma_C}{\lambda}\|u''\| + \gamma\eta \quad (11.16)$$

$$\leq e^{-\lambda\tau}V(x_\tau - x_{\tau\eta\omega}) + \frac{\sigma_C}{\lambda}\|u''\| + \gamma\eta \quad (11.17)$$

$$\leq e^{-\lambda\tau}\underline{\alpha}\varepsilon + \frac{\sigma_C}{\lambda}\omega + \gamma\eta \quad (11.18)$$

$$\leq \underline{\alpha}\varepsilon. \quad (11.19)$$

The reverse direction is similarly shown. If $(x_\tau, x_{\tau\eta\omega}) \in R_\varepsilon$ and $x_{\tau\eta\omega} \xrightarrow[\tau\eta\omega]{u_{\tau\eta\omega}} x'_{\tau\eta\omega}$, then we claim that $(x'_\tau, x'_{\tau\eta\omega}) \in R_\varepsilon$ where x'_τ is given by $x_\tau \xrightarrow[\tau]{u_{\tau\eta\omega}} x'_\tau$. The claim follows directly from inequalities (11.11) through (11.19) by using $u_\tau = u_{\tau\eta\omega}$. □

Example 11.15. To illustrate Theorem 11.12 we consider the linear control system defined by:

$$A = \begin{bmatrix} -1 & 1 \\ -8 & 5 \end{bmatrix}, \qquad C = \begin{bmatrix} 0 \\ 1 \end{bmatrix}.$$

Since the origin is not an asymptotically stable equilibrium point for $(\mathbb{R}^n, A)$, we first design the feedback control law:

$$u = Kx = 7x_1 - 6x_2 + u'$$

rendering the origin an asymptotically stable equilibrium point for the linear dynamical system $(\mathbb{R}^n, A + CK)$ where:

$$A + CK = \begin{bmatrix} -1 & 1 \\ -1 & -1 \end{bmatrix}.$$

Using the function $V(x) = \sqrt{x^T P x}$ with:

$$P = \begin{bmatrix} 1 & \frac{1}{16} \\ \frac{1}{16} & 1 \end{bmatrix}$$

as a Lyapunov function we obtain:

$$\gamma = \frac{17}{4\sqrt{15}}, \quad \lambda = \frac{16 - \sqrt{2}}{17}, \quad \underline{\alpha} = \frac{15}{16}, \quad \overline{\alpha} = \frac{17}{16}.$$

For a sampling time $\tau = 0.25$ and a precision $\varepsilon = 0.1$ we conclude from (11.6) that η needs to be smaller than 0.017. We choose $\eta = \frac{\sqrt{2}}{100} \approx 0.014$, restrict Σ to the set $[-1, 1] \times [-1, 1]$, and define $\mathcal{C}$ as the finite set consisting of constant input curves assuming values on $\{-0.5, -0.25, 0, 0.25, 0.5\}$. Although $S_{\tau\eta}(\Sigma)$ is only guaranteed to be approximately simulated by $S_\tau(\Sigma)$, several control problems that are difficult to solve directly on $S_\tau(\Sigma)$ become fairly straightforward computations on $S_{\tau\eta}(\Sigma)$. Consider the safety game for system $S_\tau(\Sigma)$ and specification set $[-0.35, -0.15] \times [-0.15, 0.15]$. It is quite difficult to solve this problem on $S_\tau(\Sigma)$, but it is immediate to solve it on $S_{\tau\eta}(\Sigma)$ due to its finite-state nature. According to the discussion in Section 11.3, if we synthesize a controller S_{cont} solving the safety game for system $S_{\tau\eta}(\Sigma)$ and specification set W, the controller $S_{cont} \times_{\mathcal{F}} S_{\tau\eta}(\Sigma)$ solves the safety game for system $S_\tau(\Sigma)$ and specification set $W^{\frac{1}{2}\varepsilon}$. Therefore, we define W as $W = [[-0.3, -0.1] \times [-0.1, 0.1]]_\eta$ and use the operator F_W defined in Chapter 6 to solve the safety game. The solution of this game is shown in Figure 11.1 where transitions with the same source and destination are not displayed to keep the figure legible.

Example 11.16. The synthesis of trajectories satisfying desired specifications can also be easily done on $S_{\tau\eta}(\Sigma)$. Assume that we are interested in designing a periodic trajectory passing through $(0.2, 0)$ and $(-0.2, 0)$. Since $S_{\tau\eta}(\Sigma)$ is finite-state, this problem reduces to a simple search on a graph. A possible solution is shown in Figure 11.2 where, in addition to the transitions of $S_{\tau\eta}(\Sigma)$, we also show several trajectories of the closed-loop system. More elaborate control problems can be solved on $S_{\tau\eta}(\Sigma)$ with similar ease by resorting to the synthesis algorithms in Chapter 6. $\triangleleft$

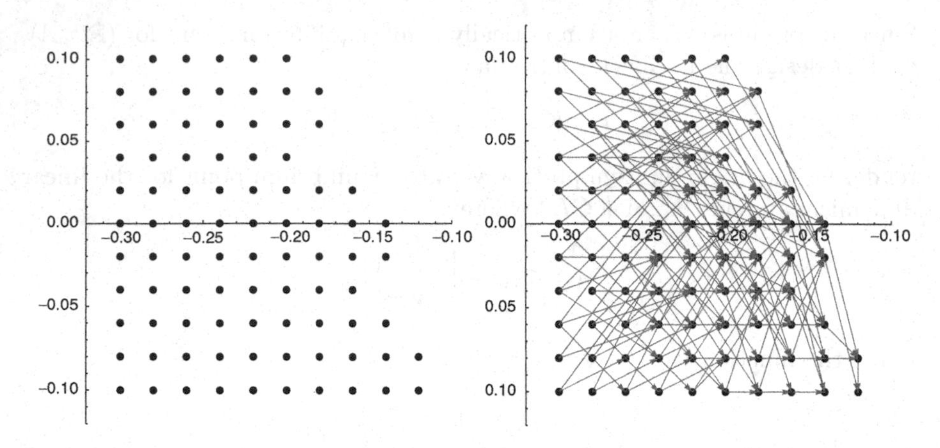

Fig. 11.1. Solution to the safety game for system $S_{\tau\eta}(\Sigma)$ and specification set $W = [[-0.3, -0.1] \times [-0.1, 0.1]]_\eta$. The left figure shows the states in W from which it is possible to control the system to remain within W. The right figure shows the corresponding transitions for which the source and destination are not the same state.

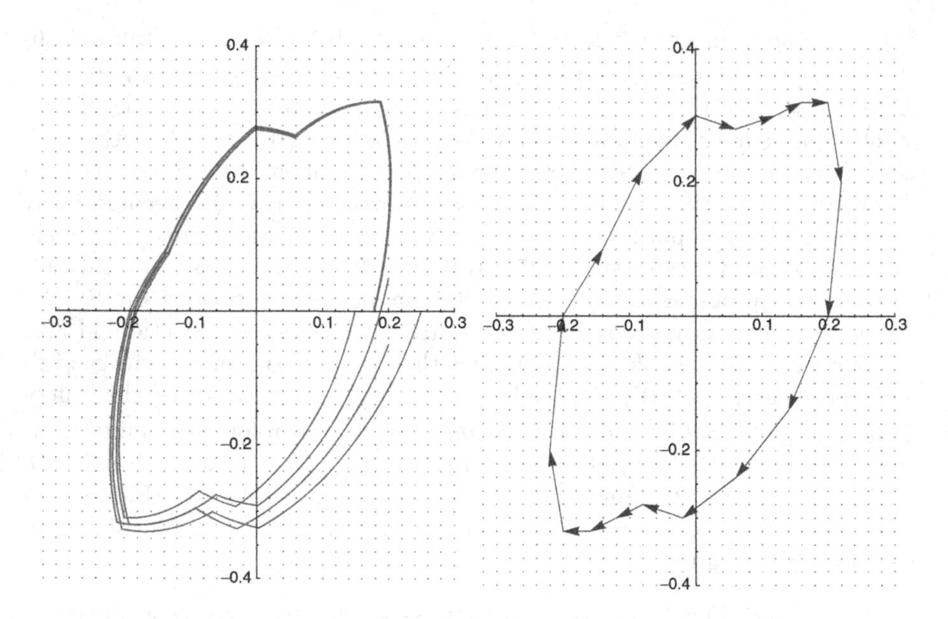

Fig. 11.2. Periodic trajectory passing through the points $(-0.2, 0)$ and $(0.2, 0)$. The left figure shows the transitions of $S_{\tau\eta}(\Sigma)$. The right figure shows several trajectories of the controlled system for different initial conditions.

When disturbance inputs are present, it is still possible to construct a finite-state system which is ε-approximate anternatingly bisimilar to $S_\tau(\Sigma)$ by a careful analysis of reachable sets. From equality (11.2) we know that all the possible contributions of the control and disturbance inputs to ξ are captured by the reachable sets:

$$\mathcal{R}_{\tau\mathcal{C}} = \left\{ x \in X_\tau : x = \int_0^\tau e^{A(\tau-t)} C\chi(t)dt,\ \chi \in \mathcal{C} \right\},$$
$$\mathcal{R}_{\tau\mathcal{D}} = \left\{ x \in X_\tau : x = \int_0^\tau e^{A(\tau-t)} D\delta(t)dt,\ \delta \in \mathcal{D} \right\}.$$

Through these sets we can indirectly quantize the inputs leading to the system $S_{\tau\eta\eta}(\Sigma)$.

Definition 11.17. *The system $S_{\tau\eta\eta} = (X_{\tau\eta\eta}, U_{\tau\eta\eta}, \underset{\tau\eta\eta}{\longrightarrow}, Y_{\tau\eta\eta}, H_{\tau\eta\eta})$ associated with an affine control system $\Sigma = (\mathbb{R}^n, \mathcal{C}, \mathcal{D}, A, C, D)$, with $\tau, \eta \in \mathbb{R}^+$, and with a set $\mathcal{D}_\eta \subseteq [\mathbb{R}^n]_\eta$ satisfying $\mathbf{d}_h(\mathcal{D}_\eta, \mathcal{R}_{\tau\mathcal{D}}) \leq \eta$, consists of:*

- $X_{\tau\eta\eta} = [\mathbb{R}^n]_\eta$;
- *$U_{\tau\eta\eta}$ is any subset of $[\mathbb{R}^m]_\eta$ satisfying $\mathbf{d}_h(U_{\tau\eta\eta}, \mathcal{R}_{\tau\mathcal{C}}) \leq \eta$;*
- *$x_{\tau\eta\eta} \xrightarrow[\tau\eta\eta]{\chi} x'_{\tau\eta\eta}$ if there exist $\chi \in U_{\tau\eta\eta}$, $\delta \in \mathcal{D}_\eta$, and a trajectory $\xi_{x_{\tau\eta\eta}00} : [0, \tau] \to \mathbb{R}^n$ of Σ satisfying:*
$$\|\xi_{x_{\tau\eta\eta}00}(\tau) + \chi + \delta - x'_{\tau\eta\eta}\| \leq \eta; \tag{11.20}$$
- $Y_{\tau\eta\eta} = \mathbb{R}^n$;
- $H_{\tau\eta\eta} = \imath : X_{\tau\eta\eta} \hookrightarrow \mathbb{R}^n$.

The construction of $S_{\tau\eta\eta}(\Sigma)$ requires the knowledge of the reachable sets $\mathcal{R}_{\tau\mathcal{C}}$ and $\mathcal{R}_{\tau\mathcal{D}}$. However, the computation of these sets does not need to be exact. Using the method described in Section 7.6, we can compute approximations $\widehat{\mathcal{R}}_{\tau\mathcal{C}}$ and $\widehat{\mathcal{R}}_{\tau\mathcal{D}}$ to the sets $\mathcal{R}_{\tau\mathcal{C}}$ and $\mathcal{R}_{\tau\mathcal{D}}$ with approximating errors $e_\mathcal{C}$ and $e_\mathcal{D}$, *i.e.*:

$$\mathbf{d}_h(\widehat{\mathcal{R}}_{\tau\mathcal{C}}, \mathcal{R}_{\tau\mathcal{C}}) \leq e_\mathcal{C} \qquad \mathbf{d}_h(\widehat{\mathcal{R}}_{\tau\mathcal{D}}, \mathcal{R}_{\tau\mathcal{D}}) \leq e_\mathcal{D}.$$

Hence, we can redefine $X_{\tau\eta\eta}$ as $[\mathbb{R}^n]_{\eta - e_\mathcal{C} - e_\mathcal{D}}$ and declare the existence of a transition $x_{\tau\eta\eta} \xrightarrow[\tau\eta\eta]{u} x'_{\tau\eta\eta}$ when $\|\xi_{x_{\tau\eta\eta}00}(\tau) + \widehat{\chi} + \widehat{\delta} - x'_{\tau\eta\eta}\| \leq \eta - e_\mathcal{C} - e_\mathcal{D}$ for some $\widehat{\chi} \in \widehat{\mathcal{R}}_{\tau\mathcal{C}}$ and $\widehat{\delta} \in \widehat{\mathcal{R}}_{\tau\mathcal{D}}$. With this new state set and transition relation we have:

$$\begin{aligned}
\|\xi_{x_{\tau\eta\eta}00}(\tau) + \chi + \delta - x'_{\tau\eta\eta}\| &= \|\xi_{x_{\tau\eta\eta}00}(\tau) + \widehat{\chi} + \widehat{\delta} - x'_{\tau\eta\eta} + \chi - \widehat{\chi} + \delta - \widehat{\delta}\| \\
&\leq \|\xi_{x_{\tau\eta\eta}00}(\tau) + \widehat{\chi} + \widehat{\delta} - x'_{\tau\eta\eta}\| \\
&\quad + \|\chi - \widehat{\chi}\| + \|\delta - \widehat{\delta}\| \\
&\leq \eta - e_\mathcal{C} - e_\mathcal{D} + e_\mathcal{C} + e_\mathcal{D} \leq \eta
\end{aligned}$$

thus maintaining the validity of the next result intact.

Theorem 11.18. *Let $\Sigma = (\mathbb{R}^n, \mathcal{C}, \mathcal{D}, A, C, D, h)$ be an affine control system and assume that the linear control system $(\mathbb{R}^n, \mathcal{C}, \mathcal{D}, A, C, D)$ admits an ISS-Lyapunov function V of the form $V(x) = \sqrt{x^T P x}$ with $P \in \mathcal{SP}(n)$. For any desired precision $\varepsilon \in \mathbb{R}^+$, for any desired time quantization τ, and for any space quantization $\eta \in \mathbb{R}^+$ satisfying:*

$$\eta \leq \min\left\{\frac{1}{3}\gamma^{-1}\underline{\alpha}\varepsilon(1 - e^{-\lambda\tau}), \overline{\alpha}^{-1}\underline{\alpha}\varepsilon\right\}, \tag{11.21}$$

the relation $R_\varepsilon \subseteq X_\tau \times X_{\tau\eta\eta}$ defined by:

$$R_\varepsilon = \{(x_\tau, x_{\tau\eta\eta}) \in X_\tau \times X_{\tau\eta\eta} \mid V(x_\tau - x_{\tau\eta\eta}) \leq \underline{\alpha}\varepsilon\} \tag{11.22}$$

is an ε-approximate alternating bisimulation relation between $S_\tau(\Sigma)$ and $S_{\tau\eta\eta}(\Sigma)$.

Proof. We first show that R_ε is an ε-approximate alternating simulation from $S_\tau(\Sigma)$ to $S_{\tau\eta\eta}(\Sigma)$. The first two requirments in Definition 9.6 are proved as in Theorem 11.12.

Regarding the third requirement, let $(x_\tau, x_{\tau\eta\eta}) \in R_\eta$ and recall that we need to show that:

$$\forall u_\tau \in U_\tau(x_\tau)\ \exists u_{\tau\eta\eta} \in U_{\tau\eta\eta}(x_{\tau\eta\eta})\ \forall x'_{\tau\eta\eta} \in \mathrm{Post}_{u_{\tau\eta\eta}}(x_{\tau\eta\eta})\ \exists x'_\tau \in \mathrm{Post}_{u_\tau}(x_\tau)$$

satisfying $(x'_\tau, x'_{\tau\eta\eta}) \in R_\varepsilon$. Choose any $u_\tau \in U_\tau(x_\tau)$ and let $u_{\tau\eta\eta}$ be any input in $U_{\tau\eta\eta}(x_{\tau\eta\eta})$ satisfying:

$$\left\| \int_0^\tau e^{A(\tau-t)} C u_\tau(t) dt - u_{\tau\eta\eta} \right\| \leq \eta. \tag{11.23}$$

Note that such input exists by definition of $U_{\tau\eta\eta}$. Let now $x'_{\tau\eta\eta}$ be any state in $\mathrm{Post}_{u_{\tau\eta\eta}}(x_{\tau\eta\eta})$. This means that:

$$\|\xi_{x_{\tau\eta\eta}00}(\tau) + u_{\tau\eta\eta} + \delta_{\tau\eta\eta} - x'_{\tau\eta\eta}\| \leq \eta \tag{11.24}$$

for some $\delta_{\tau\eta\eta} \in \mathcal{D}_\eta$. By definition of $\mathcal{D}_\eta$, there exists $\delta_\tau \in \mathcal{D}$ such that:

$$\left\| \int_0^\tau e^{A(\tau-t)} D\delta_\tau(t) dt - \delta_{\tau\eta\eta} \right\| \leq \eta. \tag{11.25}$$

We then choose x'_τ to be the element of $\mathrm{Post}_{u_\tau}(x_\tau)$ given by $x'_\tau = \xi_{x_\tau u_\tau \delta_\tau}(\tau)$ and we claim that $(x'_\tau, x'_{\tau\eta\eta}) \in R_\varepsilon$. The claim is a direct consequence of the following chain of inequalities where we used (10.6), (11.5), (11.23), (11.25), (11.22), (11.24), (11.21), and $x'' = x_\tau - x_{\tau\eta\eta}$:

$$\begin{aligned}
V(x'_\tau - x'_{\tau\eta\eta}) &\leq V\left(x'_\tau - e^{A\tau}x_{\tau\eta\eta} - \int_0^\tau e^{A(\tau-t)}\left(C\chi_\tau(t) + D\delta_\tau(t) + h\right)dt\right) \\
&\quad +\gamma\left\|x'_{\tau\eta\eta} - e^{A\tau}x_{\tau\eta\eta} - \int_0^\tau e^{A(\tau-t)}\left(C\chi_\tau(t) + D\delta_\tau(t) + h\right)dt\right\| \\
&\leq V\left(e^{A\tau}x_\tau - e^{A\tau}x_{\tau\eta\eta}\right) \\
&\quad +\gamma\left\|e^{A\tau}x_{\tau\eta\eta} + \int_0^\tau e^{A(\tau-t)}\left(C\chi_\tau(t) + D\delta_\tau(t) + h\right)dt - x'_{\tau\eta\eta}\right\| \\
&\leq V\left(\xi_{x''00}(\tau)\right) \\
&\quad +\gamma\left\|e^{A\tau}x_{\tau\eta\eta} + \int_0^\tau e^{A(\tau-t)}hdt + u_{\tau\eta\eta} + \delta_{\tau\eta\eta} - x'_{\tau\eta\eta}\right\| \\
&\quad +\gamma\left\|\int_0^\tau e^{A(\tau-t)}C\chi_\tau(t)dt - u_{\tau\eta\eta}\right\| \\
&\quad +\gamma\left\|\int_0^\tau e^{A(\tau-t)}D\delta_\tau(t)dt - \delta_{\tau\eta\eta}\right\| \\
&\leq e^{-\lambda\tau}V\left(\xi_{x''00}(0)\right) + \gamma\left\|\xi_{x_{\tau\eta\eta}00} + u_{\tau\eta\eta} + \delta_{\tau\eta\eta} - x'_{\tau\eta\eta}\right\| \\
&\quad +\gamma\eta + \gamma\eta \\
&\leq e^{-\lambda\tau}\underline{\alpha}\varepsilon + 3\gamma\eta \leq \underline{\alpha}\varepsilon.
\end{aligned}$$

The proof that R_ε^{-1} is an ε-approximate alternating simulation from $S_{\tau\eta\eta}(\Sigma)$ to $S_\tau(\Sigma)$ is similar and thus omitted. □

The previous result can also be used in the context of verification when $C = 0$. In this case, we regard the affine control system (11.1) as a closed-loop system affected by an adversarial input δ and the verification objective is to prove that a certain property holds, independently of δ.

11.5 Symbolic models for switched affine systems

The abstraction techniques developed for dynamical and control systems remarkably generalize to switched affine systems. At this point, the reader should be able to foresee how such generalization unfolds. The first step consists in quantizing the states and approximating the transitions of a switched affine system.

Definition 11.19. *The system* $S_{\tau\eta}(\Sigma) = (X_{\tau\eta}, U_{\tau\eta}, \underset{\tau\eta}{\longrightarrow}, Y_{\tau\eta}, H_{\tau\eta})$ *associated with a switched affine system* $\Sigma = (X_a, \mathbb{R}^n, \{A_{x_a}, h_{x_a}\}_{x_a \in X_a})$ *and with* $\tau, \eta \in \mathbb{R}^+$ *consists of:*

- $X_{\tau\eta} = [\mathbb{R}^n]_\eta$;
- $U_{\tau\eta} = X_a$;
- $x \underset{\tau\eta}{\xrightarrow{u_a}} x'$ *if there exists a solution* $\xi_x : [0, \tau] \to \mathbb{R}^n$ *of the affine dynamical system* $(\mathbb{R}^n, A_{u_a}, h_{u_a})$ *satisfying* $\|\xi_x(\tau) - x'\| \leq \eta$;
- $Y_{\tau\eta} = \mathbb{R}^n$;
- $H_{\tau\eta} = \imath : X_{\tau\eta} \hookrightarrow \mathbb{R}^n$.

A close analysis of the proof of Theorem 11.14 reveals that its conclusion does not depend on the particular form of the differential equation $\dot{\xi} = A\xi + C\chi + h$ but only on the inequality $\frac{\partial V}{\partial x}(Ax + Cc) \leq -\lambda V(x) + \sigma_c \|c\|$. For many affine switched systems it is possible to find a single Lyapunov function V satisfying the inequalities:

$$\frac{\partial V}{\partial x} A_{x_a} x \leq -\lambda V(x) \qquad \forall x_a \in X_a.$$

When this is the case we say that V is a *common Lyapunov function* for Σ. The arguments in the proof of Theorem 11.14 apply directly to this case and provide the following corollary.

Corollary 11.20. *Let* $\Sigma = (X_a, \mathbb{R}^n, \{A_{x_a}, h_{x_a}\}_{x_a \in X_a})$ *be a switched affine system admitting a common Lyapunov function* V *of the form* $V(x) = \sqrt{x^T P x}$ *with* $P \in \mathcal{SP}(n)$. *For any desired precision* $\varepsilon \in \mathbb{R}^+$, *for any desired time quantization* $\tau \in \mathbb{R}^+$, *and for any space quantization* $\eta \in \mathbb{R}^+$ *satisfying:*

$$\eta \leq \min \left\{ \gamma^{-1} \underline{\alpha} \varepsilon \left(1 - e^{-\lambda \tau}\right), \overline{\alpha}^{-1} \underline{\alpha} \varepsilon \right\}, \tag{11.26}$$

the relation $R_\varepsilon \subseteq X_\tau \times X_{\tau\eta}$ *defined by:*

$$R_\varepsilon = \{(x_\tau, x_{\tau\eta}) \in X_\tau \times X_{\tau\eta} \mid V(x_\tau - x_{\tau\eta}) \leq \underline{\alpha}\varepsilon\} \tag{11.27}$$

is an ε*-approximate bisimulation relation between* $S_\tau(\Sigma)$ *and* $S_{\tau\eta}(\Sigma)$.

This result can be used in two different ways. When the inputs X_a are regarded as adversarial, $S_{\tau\eta}(\Sigma)$ can be used to verify properties that hold independently of the disturbance input. When X_a is regarded as a set of control inputs, then $S_{\tau\eta}(\Sigma)$ can be used for control design.

Corollary 11.20 can also be extended to the case when there exists a Lyapunov function V_{x_a} for every linear dynamical system $(\mathbb{R}^n, A_{x_a})$. It is well known that existence of such Lyapunov functions does not imply the existence of a common Lyapunov function for Σ. In this case, a more elaborate construction is required, building upon the concept of dwell time used to study the stability properties of switched systems.

Example 11.21. We revisit the boost DC-DC converter of Chapter 1, represented in Figure 1.7. This is a switched affine system with two modes of operation corresponding to the two positions of the switch. The dynamics in mode 1 is described by (1.16) and (1.17) while the dynamics in mode 2 is described by (1.18) and (1.19). The values of the components, given in the per unit system, are:

$$C = 70, L = 3, R_C = 0.005, R_L = 0.05, v_s = 1, R_0 = 1.$$

Before proceeding with our analysis, we make the linear change of coordinates defined by :

$$\begin{bmatrix} z_1 \\ z_2 \end{bmatrix} = \begin{bmatrix} 1 & 0 \\ 0 & 5 \end{bmatrix} \begin{bmatrix} x_1 \\ x_2 \end{bmatrix}$$

to better condition the problem numerically.

The purpose of the boost DC-DC converter is to regulate the voltage across the load resistor R_0. This objective can be reformulated as the regulation of the current flowing through the inductor, which is one of the infinite state variables. In order to synthesize a controller, we regard this problem as an instance of a safety game where the specification set W contains the desired values for the current. Although safety games are difficult to solve on $S_\tau(\Sigma)$, we can use Corollary 11.20 to construct the finite-state abstraction $S_{\tau\eta}(\Sigma)$ and solve the safety game on $S_{\tau\eta}(\Sigma)$. One possible common Lyapunov function is $V(z) = \sqrt{z^T P z}$ with:

$$P = \begin{bmatrix} 1.0224 & 0.0084 \\ 0.0084 & 1.0031 \end{bmatrix}$$

and satisfying:

$$\frac{\partial V}{\partial z} A_1 z \leq -0.0139V \qquad \frac{\partial V}{\partial z} A_2 z \leq -0.0138V.$$

Therefore, $\lambda = \min\{0.0138, 0.00139\} = 0.0138$. A bound for γ can be computed using the expression in the proof of Proposition 10.5: $\gamma = 1.0256$. We select a sampling time $\tau = 0.2$ and a precision of $\varepsilon = 3$. Although this precision is not useful for practical purposes, it will keep the symbolic model $S_{\tau\eta}(\Sigma)$ small so that it can be easily visualized. From inequality (11.26) we obtain the bound $\eta \leq 0.00807$ and set $\eta = \frac{\sqrt{2}}{200} \approx 0.0071$. With these parameters we construct $S_{\tau\eta}(\Sigma)$ and consider the safety game with specification set $W = [1.2, 1.6] \times [5.6, 5.8]$ that is easily solved by iterating the operator F_W studied in Chapter 6. In Figure 11.3, the reader can find the points in W where mode 1 should be used and the points in W where mode 2 should be used. The fixed-point of F_W is displayed in Figure 11.4 and the closed-loop system $S_c \times_{\mathcal{F}} S_{\tau\eta}(\Sigma)$ is represented in the book's cover. $\triangleleft$

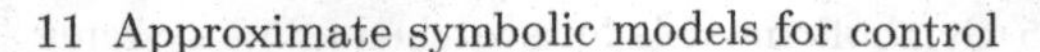

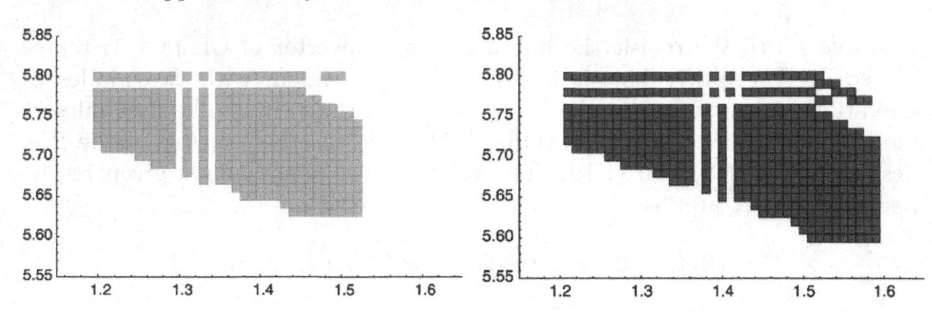

Fig. 11.3. Solution of the safety game for system $S_{\tau\eta}(\Sigma)$ and specification set $W = [1.2, 1.6] \times [5.6, 5.8]$. The points in W where mode 1 should be used are shown in the left figure and the points in W where mode 2 should be used are shown in the right figure.

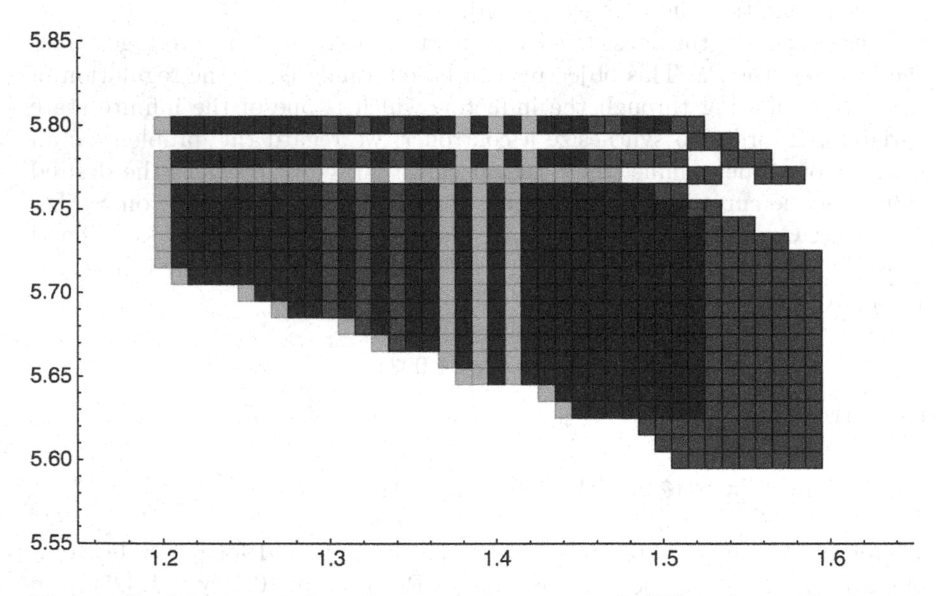

Fig. 11.4. Solution of the safety game for system $S_{\tau\eta}(\Sigma)$ and specification set $W = [1.2, 1.6] \times [5.6, 5.8]$. The fixed-point of the operator F_W is represented as the superposition of the images in Figure 11.3.

11.6 Advanced topics

In this section we show how Theorem 11.12 and Theorem 11.14 can be generalized to nonlinear control systems. The exposition will be swift and relies on advanced control theoretical concepts.

We make extensive use of comparison functions of class $\mathcal{K}$ and $\mathcal{KL}$ to simplify the arguments. A continuous function $\gamma : \mathbb{R}_0^+ \to \mathbb{R}_0^+$, is said to belong to class $\mathcal{K}$ if it is strictly increasing and $\gamma(0) = 0$; γ is said to belong to class $\mathcal{K}_\infty$

if $\gamma \in \mathcal{K}$ and $\gamma(r) \to \infty$ as $r \to \infty$. A continuous function $\beta : \mathbb{R}_0^+ \times \mathbb{R}_0^+ \to \mathbb{R}_0^+$ is said to belong to class $\mathcal{KL}$ if for each fixed s, the map $\beta(r,s)$ belongs to class $\mathcal{K}_\infty$ with respect to r and, for each fixed r, the map $\beta(r,s)$ is decreasing with respect to s and $\beta(r,s) \to 0$ as $s \to \infty$.

In a nonlinear context we need to replace the asymptotic stability assumption with the stronger assumption of incremental stability.

Definition 11.22 (Incremental global asymptotic stability). *A control system $\Sigma = (\mathbb{R}^n, \mathcal{U}, f)$ is* incrementally globally asymptotically stable (δ–GAS) *if it is forward complete and there exists a $\mathcal{KL}$ function β such that for any $t \in \mathbb{R}_0^+$, any $x, x' \in \mathbb{R}^n$, and any $\upsilon \in \mathcal{U}$, the following inequality is satisfied:*

$$\|\xi_{x\upsilon}(t) - \xi_{x'\upsilon}(t)\| \leq \beta(\|x - x'\|, t). \tag{11.28}$$

We also need the stronger notion of incremental input-to-state stability.

Definition 11.23 (Incremental global input-to-state stability). *A control system $\Sigma = (\mathbb{R}^n, \mathcal{U}, f)$ is* incrementally globally input-to-state stable (δ–ISS) *if it is forward complete and there exist a $\mathcal{KL}$ function β and a $\mathcal{K}_\infty$ function ρ such that for any $t \in \mathbb{R}_0^+$, any $x, x' \in \mathbb{R}^n$, and any $\upsilon, \upsilon' \in \mathcal{U}$, the following inequality is satisfied:*

$$\|\xi_{x\upsilon}(t) - \xi_{x'\upsilon'}(t)\| \leq \beta(\|x - x'\|, t) + \rho(\|\upsilon - \upsilon'\|). \tag{11.29}$$

It is clear that δ–ISS implies δ–GAS since (11.28) can be obtained from (11.29) by setting $\upsilon = \upsilon'$. Both δ–GAS and δ–ISS can be characterized by dissipation inequalities.

Definition 11.24 (δ–GAS Lyapunov function). *A smooth function $V : \mathbb{R}^n \times \mathbb{R}^n \to \mathbb{R}$ is called a* δ–GAS Lyapunov function *for a control system $\Sigma = (\mathbb{R}^n, \mathcal{U}, f)$, if there exist $\lambda \in \mathbb{R}^+$ and $\mathcal{K}_\infty$ functions $\underline{\alpha}$ and $\overline{\alpha}$ such that for any $x, x' \in \mathbb{R}^n$ and any $u \in \mathbb{R}^m$ we have:*

$$\underline{\alpha}(\|x - x'\|) \leq V(x, x') \leq \overline{\alpha}(\|x - x'\|)$$
$$\frac{\partial V}{\partial x} f(x, u) + \frac{\partial V}{\partial x'} f(x', u) \leq -\lambda V(x, x').$$

Function V is called a δ–ISS Lyapunov function for Σ, if there exist $\mathcal{K}_\infty$ functions $\underline{\alpha}$, $\overline{\alpha}$, and σ such that for any $x, x' \in \mathbb{R}^n$ and any $u, u' \in \mathbb{R}^m$ we have:

$$\underline{\alpha}(\|x - x'\|) \leq V(x, x') \leq \overline{\alpha}(\|x - x'\|)$$
$$\frac{\partial V}{\partial x} f(x, u) + \frac{\partial V}{\partial x'} f(x', u') \leq -\lambda V(x, x') + \sigma(\|u - u'\|).$$

The following result completely characterizes δ–GAS and δ–ISS in terms of existence of Lyapunov functions.

Theorem 11.25. *For any control system $\Sigma = (\mathbb{R}^n, \mathcal{U}, f)$ the following holds:*

1. *if the elements of $\mathcal{U}$ assume values on compact set $K \subseteq \mathbb{R}^m$, then Σ is δ–GAS if and only if it admits a δ–GAS Lyapunov function;*
2. *if the elements of $\mathcal{U}$ assume values on closed and convex set $K \subseteq \mathbb{R}^m$ containing the origin, and if $f(0,0) = 0$, then Σ is δ–ISS if it admits a δ–ISS Lyapunov function. Moreover if the elements of $\mathcal{U}$ assume values on compact set $K \subseteq \mathbb{R}^m$, existence of a δ–ISS Lyapunov function is equivalent to δ–ISS.*

Theorems 11.12 and 11.14 can now be generalized to the nonlinear context.

Theorem 11.26. *Let $\Sigma = (\mathbb{R}^n, \mathcal{U}, f)$ be a control system admitting a δ-GAS Lyapunov function V satisfying:*

$$V(x, x') - V(x, x'') \leq \gamma(\|x' - x''\|)$$

for some class $\mathcal{K}_\infty$ function γ and for every $x, x', x'' \in \mathbb{R}^n$. For any desired precision $\varepsilon \in \mathbb{R}^+$, for any desired time quantization $\tau \in \mathbb{R}^+$, and for any space quantization $\eta \in \mathbb{R}^+$ satisfying:

$$\eta \leq \min\left\{\gamma^{-1}\left((1 - e^{\lambda\tau})\overline{\alpha}(\varepsilon)\right), \overline{\alpha}^{-1} \circ \underline{\alpha}(\varepsilon)\right\}, \tag{11.30}$$

the relation $R_\varepsilon \subseteq X_{\tau\eta} \times X_\tau$ defined by:

$$R_\varepsilon = \{(x_{\tau\eta}, x_\tau) \in X_{\tau\eta} \times X_\tau \mid V(x_\tau, x_{\tau\eta}) \leq \underline{\alpha}(\varepsilon)\} \tag{11.31}$$

is a surjective ε-approximate simulation relation from $S_{\tau\eta}(\Sigma)$ to $S_\tau(\Sigma)$. Moreover, if V is a δ-ISS Lyapunov function and $\mathcal{U}$ contains only constant curves, then for any desired precision $\varepsilon \in \mathbb{R}^+$, for any desired time quantization $\tau \in \mathbb{R}^+$, for any desired input quantization $\omega \in \mathbb{R}^+$, and for any space quantization $\eta \in \mathbb{R}^+$ satisfying:

$$\eta \leq \min\left\{\gamma^{-1}\left(\underline{\alpha}(\varepsilon)(1 - e^{-\lambda\tau}) - \lambda^{-1}\sigma\omega\right), \overline{\alpha}^{-1} \circ \underline{\alpha}(\varepsilon)\right\}, \tag{11.32}$$

the relation (11.31) is an ε-approximate bisimulation relation between $S_{\tau\eta\omega}(\Sigma)$ and $S_\tau(\Sigma)$.

Proof. The proof parallels the proof of Theorems 11.12 and 11.14. The only modification is the replacement of the sequence of inequalities used to prove the third condition in Definitions 9.5 and 9.6. We only provide the details for the inequalities (11.11) through (11.19) since the same argument applies to the remaining ones.

$$\begin{aligned} V(x'_\tau, x'_{\tau\eta\omega}) &= V(x'_\tau, \xi_{x_{\tau\eta\omega}u_{\tau\eta\omega}}(\tau)) + \gamma(\|\xi_{x_{\tau\eta\omega}u_{\tau\eta\omega}}(\tau) - x'_{\tau\eta\omega}\|) \\ &\leq V(\xi_{x_\tau u_\tau}(\tau), \xi_{x_{\tau\eta\omega}u_{\tau\eta\omega}}(\tau)) + \gamma(\eta) \\ &\leq e^{-\lambda\tau}V(x_\tau, x_{\tau\eta\omega}) + \frac{\sigma}{\lambda}\omega + \gamma(\eta) \\ &\leq e^{-\lambda\tau}\underline{\alpha}(\varepsilon) + \frac{\sigma}{\lambda}\omega + \gamma(\eta) \leq \underline{\alpha}(\varepsilon). \end{aligned}$$

□

11.7 Notes

The results in this chapter are quite recent and based on [PGT08, GPT09, PT09]. Earlier work relating stability properties of control systems to the existence of approximate simulation relations appeared in [Tab06, Tab08a]. In [GPT09], the reader can find a nonlinear version of Corollary 11.20 that does not require a common Lyapunov function. Instead, it relies on the concept of dwell time from the switched systems literature. The generalization of Theorem 11.18 to nonlinear systems is reported in [PT09].

The boost DC-DC example is taken from [GPM04] and was also used in [GPT09]. In this reference, the interested readers can find a more detailed treatment of Example 11.21.

The discussion of δ-ISS properties in Section 11.6 follows [Ang02] where the proof of Theorem 11.25 can be found.

Although we only used the notions of approximate simulation and bisimulation to construct finite-state abstractions, they can also be used to construct infinite-state abstractions to simplify controller design problems [GP09].

Controller synthesis based on finite-state approximate models had already been discussed in [RO98, MRO02] although the notion of approximation used in these references corresponds that what we defined as a simulation relation.

As mentioned in Section 10.5, the abstraction techniques discussed in Part IV have not yet been extended to hybrid systems. The main difficulty consists in inferring, from the entrance of a single trajectory in a guard set, the entrance of the surrounding trajectories in the same guard set . The exception of switched systems, discussed in Section 11.5, is easy to explain since for this class of hybrid systems the guards coincide with the invariant sets. A very recent and promising research direction that may lead to the desired extension is a direct study of the stability properties of hybrid systems and its corresponding Lyapunov functions [CTG07, CGT08].

11.7 Notes

Appendix

We review Tarski's fixed-point theorem, see [Tar55], and some of its corollaries.

Notation

In this appendix we denote the image of a set $K \subseteq Z$ under a function $f : Z \to W$ by $f(K) = \{w \in W \mid w = f(k) \text{ for some } k \in K\}$.

A.1 Lattice theory

We start with a relation $R \subseteq X \times X$ on a set X. Relation R is said to be: *reflexive* when $(x, x) \in R$ for every $x \in X$; *anti-symmetric* when $(x, x') \in R$ and $(x', x) \in R$ imply $x = x'$; and *transitive* when $(x, x') \in R$ and $(x', x'') \in R$ imply $(x, x'') \in R$. When $f : Z \to Z$, we denote by f^i the i-fold composition of f with itself.

Given a relation $R \subseteq X \times X$, we denote a pair $(x, x') \in R$ by xRx'.

Definition A.1 (Pre-order, partial order, and total order). *A* pre-order *on a set X, denoted by $\sqsubseteq$, is a relation on X that is reflexive and transitive. A pre-order $\sqsubseteq$ is said to be a* partial order *when $\sqsubseteq$ is also anti-symmetric. A partial order is said to be* total *when for every $x, x' \in X$ either $x \sqsubseteq x'$ or $x' \sqsubseteq x$ holds. The pair $(X, \sqsubseteq)$, where $\sqsubseteq$ is a partial order on X, is called a partially ordered set.*

Consider now a subset $X' \subseteq X$ and an element $x \in X$. The element x is said to be the *supremum* of X', denoted by $\sup X'$, when the following two conditions hold:

1. $\forall x' \in X' \; x' \sqsubseteq x$ (x is an upper bound);
2. $\forall x'' \in X \; (\forall x' \in X' \;\; x' \sqsubseteq x'') \implies x \sqsubseteq x''$ (x is the smallest upper bound).

Dualizing the above definition we obtain the notion of *infimum*. The element $x \in X$ is the infimum of $X' \subseteq X$, denoted by $\inf X'$, if the following two conditions hold:

1. $\forall x' \in X' \; x \sqsubseteq x'$ (x is a lower bound);
2. $\forall x'' \in X \; (\forall x' \in X' \quad x'' \sqsubseteq x') \implies x'' \sqsubseteq x$ (x is the greatest lower bound).

Definition A.2 (Lattice and complete lattice). *A partially ordered set* $(X, \sqsubseteq)$ *is said to be a* lattice *if for any finite set* $X' \subseteq X$ *the supremum of* X' *and the infimum of* X' *exist and belong to* X*. Whenever* $\sup X'$ *and* $\inf X'$ *exist and belong to* X *for any set* $X' \subseteq X$ *we say that* $(X, \sqsubseteq)$ *is a complete lattice.*

The reader should check that $(2^Z, \subseteq)$ is a complete lattice for any set Z, for the set of all subsets of Z, denoted by 2^Z, and for the usual set inclusion relation $\subseteq$. In this lattice, $\sup\{A, B\} = A \cup B$ and $\inf\{A, B\} = A \cap B$ for any $A, B \in 2^Z$.

A totally ordered subset $X' \subseteq X$ is said to be a *chain*. A chain $\{x_i\}_{i \in \mathbb{N}}$, where $x_i \in X'$, is *increasing* when $x_i \sqsubseteq x_{i+1}$ and *decreasing* when $x_{i+1} \sqsubseteq x_i$ for all $i \in \mathbb{N}$.

Definition A.3. *Let* $(X, \sqsubseteq)$ *be a complete lattice and consider a function* $f : X \to X$*. Function* f *is said to be:*

- monotone *if* $x \sqsubseteq x' \implies f(x) \sqsubseteq f(x')$ *for every* $x, x' \in X$*;*
- sup-continuous *if* $f(\sup\{x_i\}_{i \in \mathbb{N}}) = \sup\{f(x_i)\}_{i \in \mathbb{N}}$ *for every increasing chain* $\{x_i\}_{i \in \mathbb{N}}$ *with* $x_i \in X$*;*
- inf-continuous *if* $f(\inf\{x_i\}_{i \in \mathbb{N}}) = \inf\{f(x_i)\}_{i \in \mathbb{N}}$ *for every decreasing chain* $\{x_i\}_{i \in \mathbb{N}}$ *with* $x_i \in X$*.*

A.2 fixed-points

A fixed-point of a function $f : X \to X$ is an element $x \in X$ satisfying $f(x) = x$. Tarski's fixed-point theorem provides sufficient conditions for the supremum and the infimum of the set of fixed-points of a function to be fixed-points.

Theorem A.4. *Let* $(X, \sqsubseteq)$ *be a complete lattice, let* $f : X \to X$ *be a function, and denote by* $Y = \{x \in X \mid f(x) = x\}$ *the set of all fixed-points of* f*. If* f *is monotone then the following holds true:*

- $\sup Y \in Y$ *and* $\sup Y = \{x \in X \mid x \sqsubseteq f(x)\}$*;*
- $\inf Y \in Y$ *and* $\inf Y = \{x \in X \mid f(x) \sqsubseteq x\}$*.*

Although Tarski's theorem asserts that a monotone function always has an infimal and a supremal fixed-point, it does not state how such fixed-points can be computed. The next result addresses this question.

Theorem A.5. *Let $(X, \sqsubseteq)$ be a complete lattice, let $f : X \to X$ be a function, and denote by $Y = \{x \in X \mid f(x) = x\}$ the set of all fixed-points of f. If f is* sup*-continuous, then:*

$$\inf Y = \sup \{\inf X, f(\inf X), f^2(\inf X), \ldots\}. \tag{12.1}$$

Dually, If f is inf*-continuous, then:*

$$\sup Y = \inf \{\sup X, f(\sup X), f^2(\sup X), \ldots\}. \tag{12.2}$$

When the set X is finite any increasing or decreasing chain $\{x_i\}_{i \in \mathbb{N}}$ is necessarily finite in the sense that there exists a $k \in \mathbb{N}$ such that $x_k = x_j$ for $j \geq k$. This implies that the supremum of an increasing chain is:

$$\sup\{x_i\}_{i=1,2,\ldots,k} = x_k = \lim_{i \to \infty} x_i.$$

If we apply a monotone function $f : X \to X$ to the increasing chain $\{x_i\}_{i \in \mathbb{N}}$ we obtain the increasing chain:

$$f(x_1) \sqsubseteq f(x_2) \sqsubseteq f(x_3) \sqsubseteq \ldots \sqsubseteq f(x_k) = f(x_{k+1}) = f(x_{k+2}) = \ldots$$

and its supremum is:

$$\sup\{f(x_i)\}_{i \in \mathbb{N}} = f(x_k) = \lim_{i \to \infty} f(x_i).$$

This shows that any monotone function with respect to a lattice defined on a finite set is a sup-continuous function. The same argument also shows that f is inf-continuous.

Consider now equality (12.1). By definition of infimum, $\inf X \sqsubseteq X'$ for any $X' \subseteq X$. Therefore, $\inf X \sqsubseteq f(\inf X)$. Since f is monotone, $\inf X \sqsubseteq f(\inf X)$ implies $f(\inf X) \sqsubseteq f^2(\inf X)$ and by induction $f^i(\inf X) \sqsubseteq f^{i+1}(\inf X)$ for every $i \in \mathbb{N}$. We conclude that $\{f^i(\inf X)\}_{i \in \mathbb{N}_0}$ is an increasing chain. Finiteness of X now ensures that this chain is finite and:

$$\sup \{\inf X, f(\inf X), f^2(\inf X), \ldots\} = \lim_{i \to \infty} f^i(\inf X).$$

Dually, we have:

$$\inf \{\sup X, f(\sup X), f^2(\sup X), \ldots\} = \lim_{i \to \infty} f^i(\sup X).$$

We summarize this discussion in the next result.

Corollary A.6. *Let $(X, \sqsubseteq)$ be a complete lattice over a finite set X, let $f : X \to X$ be a function, and denote by $Y = \{x \in X \mid f(x) = x\}$ the set of all fixed-points of f. If f is monotone then:*

$$\inf Y = \lim_{i \to \infty} f^i(\inf X) \qquad \sup Y = \lim_{i \to \infty} f^i(\sup X).$$

References

ABDM00. A. Asarin, O. Bournez, T. Dang, and O. Maler. Approximate reachability analysis of piecewise-linear dynamical systems. In *Hybrid Systems: Computation and Control*, volume 1790 of *Lecture Notes in Computer Science*, pages 20–31. Springer, 2000.

ACH+95. R. Alur, C. Courcoubetis, N. Halbwachs, T. A. Henzinger, P.-H. Ho, X. N., A. Olivero, J. Sifakis, and S. Yovine. The algorithmic analysis of hybrid systems. *Theoretical Computer Science*, 138:3–34, 1995.

AD90. R. Alur and D. L. Dill. Automata for modeling real-time systems. In *Proceedings of the 17th International Colloquium on Automata, Languages and Programming, ICALP*, volume 443 of *Lecture Notes in Computer Science*, pages 322–335. Springer, 1990.

AD94. R. Alur and D. L. Dill. A theory of timed automata. *Theoretical Computer Science*, 126(2):183–235, 1994.

AHKV98. R. Alur, T. Henzinger, O. Kupferman, and M. Vardi. Alternating refinement relations. In *Proceedings of the 8th International Conference on Concurrence Theory*, number 1466 in Lecture Notes in Computer Science, pages 163–178. Springer, 1998.

Alu99. R. Alur. Timed automata. In *11th International Conference on Computer-Aided Verification*, volume 1633 of *Lecture Notes in Computer Sience*, pages 8–22. Springer, 1999.

AM97. P. J. Antsaklis and A. N. Michel. *Linear Systems*. McGraw-Hill, 1997.

AM04. R. Alur and P. Madhusudan. Decision problems for timed automata: A survey. In *International School on Formal Methods for the Design of Computer, Communication and Software Systems, SFM-RT 2004*, volume 3185 of *Lecture Notes in Computer Science*, pages 1–24. Springer, 2004.

AMP95. E. Asarin, O. Maler, and A. Pnueli. Symbolic controller synthesis for discrete and timed systems. In *Hybrid Systems II*, volume 999 of *Lecture Notes in Computer Science*, pages 1–20. Springer, 1995.

AMR88. R. Abraham, J. Marsden, and T. Ratiu. *Manifolds, Tensor Analysis and Applications*. Applied Mathematical Sciences. Springer, 1988.

Ang02. D. Angeli. A Lyapunov approach to incremental stability properties. *IEEE Transactions on Automatic Control*, 47(3):410–421, 2002.

AS87. B. Alpern and F. B. Schneider. Recognizing safety and liveness. *Distributed Computing*, 2(3):117–126, 1987.

AVW03. A. Arnold, A. Vincent, and I. Walukiewicz. Games for synthesis of controllers with partial observation. *Theoretical Computer Science*, 28(1):7–34, 2003.

BBF+01. B. Berard, M. Bidoit, A. Finkel, F. Laroussinie, A. Petit, L. Petrucci, and P. Schnoebelen. *Systems and Software Verification*. Springer, 2001.

BH06. C. Belta and L.C.G.J.M. Habets. Controlling a class of nonlinear systems on rectangles. *IEEE Transactions on Automatic Control*, 51(11):1749–1759, 2006.

BM05. T. Brihaye and C. Michaux. On the expressiveness and decidability of o-minimal hybrid systems. *Journal of Complexity*, 21(4):447–478, 2005.

BMP02. A. Bicchi, A. Marigo, and B. Piccoli. On the reachability of quantized control systems. *IEEE Transactions on Automatic Control*, 47(4):546–563, 2002.

BT00. O. Botchkarev and S. Tripakis. Verification of hybrid systems with linear differential inclusions using ellipsoidal approximations. In *Hybrid Systems: Computation and Control*, volume 1790 of *Lecture Notes in Computer Science*, pages 73–88. Springer, 2000.

BY04. J. Bengtsson and W. Yi. Timed automata: Semantics, algorithms and tools. In *Lecture Notes on Concurrency and Petri Nets*, volume 3098 of *Lecture Notes in Computer Science*, pages 87–124. Springer, 2004.

CGP99. E. M. Clarke, O. Grumberg, and D. Peled. *Model Checking*. MIT Press, 1999.

CGT08. C. Chaohong, R. Goebel, and A. Teel. Smooth Lyapunov functions for hybrid systems Part II: (Pre)Asymptotically stable compact sets. *IEEE Transactions on Automatic Control*, 53(3):734–748, 2008.

CK01. A. Chotinan and B.H. Krogh. Verification of infinite state dynamical systems using approximate quotient transition systems. *IEEE Transactions on Automatic Control*, 46(9):1401–1410, 2001.

CK03. A. Chutinan and B. Krogh. Computational techniques for hybrid system verification. *IEEE Transactions on Automatic Control*, 48(1):64–75, 2003.

CL99. C. Cassandras and S. Lafortune. *Introduction to discrete event systems*. Kluwer Academic Publishers, Boston, MA, 1999.

CTG07. C. Chaohong, A. Teel, and R. Goebel. Smooth Lyapunov functions for hybrid systemsPart I: Existence is equivalent to robustness. *IEEE Transactions on Automatic Control*, 52(7):1264–1277, 2007.

dAFS04. L. de Alfaro, M. Faella, and M. Stoelinga. Linear and branching metrics for quantitative transition systems. In *Proceedings of the 31st International Colloquium on Automata, Languages and Programming, ICALP*, volume 3142 of *Lecture Notes in Computer Science*, pages 97–109. Springer, 2004.

dAHM01. L. de Alfaro, T. A. Henzinger, and R. Majumdar. Symbolic algorithms for infinite-state games. In *Proceedings of the 12th International Conference on Concurrency Theory (CONCUR)*, volume 2154 of *Lecture Notes in Computer Science*, pages 536–550. Springer, 2001.

DGJP99. J. Desharnais, V. Gupta, R. Jagadeesan, and P. Panangaden. Metrics for labeled markov systems. In *Proceedings of 10th International Confer-*

ence on Concurrency Theory, volume 1664 of *Lecture Notes in Computer Science*, pages 258–273. Springer, 1999.

FDF05. E. Frazzoli, M. A. Dahleh, and E. Feron. Maneuver-based motion planning for nonlinear systems with symmetries. *IEEE Transactions on Robotics*, 21(6):1077–1091, 2005.

FGP06. G. E. Fainekos, A. Girard, and G. J. Pappas. Temporal logic verification using simulation. In *Formal Modelling and Analysis of Timed Systems*, volume 4202 of *Lecture Notes in Computer Science*, pages 171–186. Springer, 2006.

FKPY07. E. Fersman, P. Krcal, P. Pettersson, and W. Yi. Task automata: Schedulability, decidability and undecidability. *Information and Computation*, 205(8):1149–1172, 2007.

GGM06. A. Girard, C. Le Guernic, and O. Maler. Efficient computation of reachable sets of linear time-invariant systems with inputs. In *Hybrid Systems: Computation and Control*, volume 3927 of *Lecture Notes in Computer Science*, pages 257–271. Springer, 2006.

GHJ97. V. Gupta, T.A. Henzinger, and R. Jagadeesan. Robust timed automata. In *Proceedings of the International Workshop on Hybrid and Real-Time Systems*, volume 1201 of *Lecture Notes in Computer Science*, pages 331–345. Springer, 1997.

Gir05. A. Girard. Reachability of uncertain linear systems using zonotopes. In *Hybrid Systems: Computation and Control*, volume 3414 of *Lecture Notes in Computer Science*, pages 291–305. Springer, 2005.

Gir07. A. Girard. Approximately bisimilar finite abstractions of stable linear systems. In *Hybrid Systems: Computation and Control*, volume 4416 of *Lecture Notes in Computer Science*, pages 231–244. Springer, 2007.

GP05. A. Girard and G. Pappas. Approximate bisimulations for nonlinear dynamical systems. In *Proceedings of the 44th IEEE Conference on Decision and Control*, pages 684–689, Seville, Spain, 2005.

GP06. A. Girard and G. J. Pappas. Verification using simulation. In *Hybrid Systems: Computation and Control*, volume 3927 of *Lecture Notes in Computer Science*, pages 272–286. Springer, 2006.

GP07. A. Girard and G. J. Pappas. Approximation metrics for discrete and continuous systems. *IEEE Transactions on Automatic Control*, 52(5):782–798, 2007.

GP09. A. Girard and G.J. Pappas. Hierarchical control system design using approximate simulation. *Automatica*, 45(2):566–571, 2009.

GPM04. T. Geyer, G. Papafotiou, and M. Morari. On the optimal control of switch-mode dc-dc converters. In *Hybrid Systems: Computation and Control*, volume 2993 of *Lecture Notes in Computer Sience*, pages 342–356. Springer, 2004.

GPT09. A. Girard, G. Pola, and P. Tabuada. Approximately bisimilar symbolic models for incrementally stable switched systems. *IEEE Transactions on Automatic Control*, 2009. In press.

HCvS06. L. C. G. J. M. Habets, P.J. Collins, and J. H. van Schuppen. Reachability and control synthesis for piecewise-affine hybrid systems on simplices. *IEEE Transactions on Automatic Control*, 51(6):938–948, 2006.

Hen96. T.A. Henzinger. The theory of hybrid automata. In *Proceedings of the 11th Annual IEEE Symposium on Logic in Computer Science*, pages 278–292. IEEE Computer Society Press, 1996.

HKPV98. T. A. Henzinger, P. W. Kopke, A. Puri, and P. Varaiya. What's decidable about hybrid automata? *Journal of Computer and System Sciences*, 57:94–124, 1998.

HMP05. T.A. Henzinger, R. Majumdar, and V.S. Prabhu. Quantifying similarities between timed systems. In *FORMATS: Formal Modeling and Analysis of Timed Systems*, Lecture Notes in Computer Science 3829, pages 226–241. Springer, 2005.

HvS01. L.C.G.J.M. Habets and J. H. van Schuppen. Control of piecewise-linear hybrid systems on simplices and rectangles. In *Hybrid Systems: Computation and Control*, volume 2034 of *Lecture Notes in Computer Sience*, pages 261–274. Springer, 2001.

HvS04. L. C. G. J. M. Habets and J. H. van Schuppen. A control problem for affine dynamical systems on a full-dimensional polytope. *Automatica*, 40(1):21–35, 2004.

KA03. X. Koutsoukos and P. Antsaklis. Safety and reachability of piecewise linear hybrid dynamical systems based on discrete abstractions. *Journal of Discrete Event Dynamic Systems: Theory and Applications*, 13(3):203–243, 2003.

KG95. R. Kumar and V.K. Garg. *Modeling and Control of Logical Discrete Event Systems*. Kluwer Academic Publishers, 1995.

KV00. A. Kurzhanski and P. Varaiya. Ellipsoidal techniques for reachability analysis. In *Hybrid Systems: Computation and Control*, volume 1790 of *Lecture Notes in Computer Science*, pages 202–214. Springer, 2000.

LM86. J. Levine and R. Marino. Nonlinear system immersion, observers and finite-dimensional filters. *Systems and Control Letters*, 7(2):133–142, 1986.

LPS99. G. Lafferriere, G. J. Pappas, and S. Sastry. Hybrid systems with finite bisimulations. In *Hybrid Systems V*, volume 1567 of *Lecture Notes in Computer Science*, pages 186–203. Springer, 1999.

LPS00. G. Lafferriere, G. J. Pappas, and S. Sastry. O-minimal hybrid systems. *Mathematics of Control, Signals and Systems*, 13(1):1–21, 2000.

Lun94. J. Lunze. Qualitative modeling of linear dynamical systems with quantized state measurements. *Automatica*, 30(3):417–431, 1994.

Mil89. R. Milner. *Communication and Concurrency*. International Series in Computer Science. Prentice Hall, 1989.

MRO02. T. Moor, J. Raisch, and S. O'Young. Discrete supervisory control of hybrid systems based on l-complete approximations. *Journal of Discrete Event Dynamic Systems*, 12(1):83–107, 2002.

MT02. P. Madhusudan and P.S. Thiagarajan. Branching time controllers for discrete event systems. *Theoretical Computer Science*, 274:117–149, 2002.

Ner58. A. Nerode. Linear automaton transformations. *Proceedings of the American Mathematical Society*, 9:541–544, 1958.

Par81. D. Park. Concurrency and automata on infinite sequences. In *Theoretical Computer Science*, pages 167–183, 1981.

PC08. A. Platzer and E. M. Clarke. Computing differential invariants of hybrid systems as fixed points. In *Computer Aided Verification, CAV 2008*, volume 5123 of *Lecture Notes in Computer Sience*, pages 176–189. Springer, 2008.

PGT08. G. Pola, A. Girard, and P. Tabuada. Approximately bisimilar symbolic models for nonlinear control systems. *Automatica*, 44(10):2508–2516, 2008.

PJP07. A. Prajna, A. Jadbabaie, and G.J. Pappas. A framework for worst-case and stochastic safety verification using barrier certificates. *IEEE Transactions on Automatic Control*, 52(8):1415–1428, 2007.

PPP02. S. Prajna, A. Papachristodoulou, and P. A. Parrilo. Introducing SOS-TOOLS: A general purpose sum of squares programming solver. In *Proceedings of the 41st IEEE Conference on Decision and Control*, pages 741–746, 2002.

PR89a. A. Pnueli and R. Rosner. On the synthesis of a reactive module. In *Proceedings of the 16th ACM Symposium on Principles of Programming Languages*, pages 170–190. ACM, 1989.

PR89b. A. Pnueli and R. Rosner. On the synthesis of an asynchronous reactive module. In *Proceedings of the 16th International Colloquium on Automata, Languages and Programming, ICALP*, volume 372 of *Lecture Notes in Computer Science*, pages 652–671. Springer, 1989.

PR07. S. Prajna and A. Rantzer. Convex programs for temporal verification of nonlinear dynamical systems. *SIAM Journal of Control and Optimization*, 46(3):999–1021, 2007.

PT09. G. Pola and P. Tabuada. Symbolic models for nonlinear control systems: Alternating approximate bisimulations. *SIAM Journal of Control and Optimization*, 48(2):719–733, 2009.

PVB95. A. Puri, P. Varaiya, and V. Borkar. ε-approximation of differential inclusions. In *Proceedings of the 34th IEEE Conference on Decision and Control*, pages 2892–2897, 1995.

PW97. J. W. Polderman and J.C. Willems. *Introduction to Mathematical Systems Theory: A Behavioral Approach.* Springer, New York, 1997.

QL91. H. Qin and P. Lewis. Factorization of finite state machines under strong and observational equivalences. *Formal Aspects of Computing*, 3(3):284–307, 1991.

RB06. B. Roszak and M.E. Broucke. Necessary and sufficient conditions for reachability on a simplex. *Automatica*, 42(11):1913–1918, 2006.

RB07. B. Roszak and M.E. Broucke. Reachability of a set of facets for linear affine systems with n-1 inputs. *IEEE Transactions on Automatic Control*, 52(2):359–364, 2007.

RO98. J. Raisch and S. O'Young. Discrete approximation and supervisory control of continuous systems. *IEEE Transactions on Automatic Control*, 43(4):569–573, 1998.

RW87. P.J. Ramadge and W. M. Wonham. Supervisory control of a class of discrete event systems. *SIAM Journal on Control and Optimization*, 25(1):206–230, 1987.

RW89. P.J. Ramadge and W. M. Wonham. The control of discrete event systems. *Proceedings of IEEE*, 77(1):81–98, 1989.

Sam39. P. Samuelson. Interactions between the multiplier analysis and the principle of acceleration. *The Review of Economic Statistics*, 21(2):75–78, 1939.

SKA01. J. Stiver, X. Koutsoukos, and P. Antsaklis. An invariant based approach to the design of hybrid control systems. *International Journal of Robust and Nonlinear Control*, 11(5):453–478, 2001.

Spe99. P. Speissegger. The pfaffian closure of an o-minimal structure. *J. Reine Angew. Math.*, 508:189–211, 1999.

SSM06. S. Sankaranarayanan, H. B. Sipma, and Z. Manna. Fixed point iteration for computing the time elapse operator. In *Hybrid Systems: Computation and Control*, volume 3927 of *Lecture Notes in Computer Sience*, pages 537–551. Springer, 2006.

Tab04. P. Tabuada. Open maps, alternating simulations and controller synthesis. In *Proceedings of the 15th International Conference on Concurrence Theory*, volume 3170 of *Lecture Notes in Computer Science*, pages 466–480. Springer, 2004.

Tab05. P. Tabuada. Symbolic sub-systems and symbolic control of linear systems. In *Proceedings of the 44th IEEE Conference on Decision and Control*, pages 18–23, Seville, Spain, 2005.

Tab06. P. Tabuada. Symbolic control of linear systems based on symbolic subsystems. *IEEE Transactions on Automatic Control, Special issue on symbolic methods for complex control systems*, 51(6):1003–1013, 2006.

Tab08a. P. Tabuada. An approximate simulation approach to symbolic control. *IEEE Transactions on Automatic Control*, 53(6):1406–1418, 2008.

Tab08b. P. Tabuada. Controller synthesis for bisimulation equivalence. *Systems and Control Letters*, 57(6):443–452, 2008.

Tar55. A. Tarski. A lattice theoretical fixed point and its applications. *Pacific Journal of Mathematics*, 5(2):285–309, 1955.

Tiw08. A. Tiwari. Abstractions for hybrid systems. *Formal Methods in Systems Design*, 32:57–83, 2008.

TMBO03. C. Tomlin, I. Mitchell, A. Bayen, and M. Oishi. Computational techniques for the verification of hybrid systems. *Proceedings of the IEEE*, 91(7):986–1001, 2003.

TP03. P. Tabuada and G. J. Pappas. Model checking LTL over controllable linear systems is decidable. In *Hybrid Systems: Computation and Control*, volume 2623 of *Lecture Notes in Computer Sience*, pages 498–513. Springer, 2003.

TP06. P. Tabuada and G. J. Pappas. Linear Time Logic control of discrete-time linear systems. *IEEE Transactions on Automatic Control*, 51(12):1862–1877, 2006.

vB98. F. van Breugel. *Comparative Metric Semantics of Programming Languages: Non-determinism and Recursion.* Birkhauser, Boston, 1998.

vdD98. L. van den Dries. *Tame Topology and o-minimal structures*, volume 248 of *London Mathematical Society Lecture Note Series.* Cambridge University Press, 1998.

vdE94. A. van den Essen. Locally finite and locally nilpotent derivations with applications to polynomial flows, morphisms, and $\mathcal{G}_a$-actions II. In *Proceedings of the American Mathematical Society*, volume 121, pages 667–678. American Mathematical Society, 1994.

Wan68. P.K.C. Wang. A method for approximating dynamical processes by finite-state systems. *International Journal of Control*, 8(3):285–296, 1968.

ZKJ06. C. Zhou, R. Kumar, and S. Jiang. Control of nondeterministic discrete-event systems for bisimulation equivalence. *IEEE Transactions on Automatic Control*, 51(5):754–765, 2006.

Index